COLLOIDAL DISPERSIONS

Special Publication No 43

Colloidal Dispersions

The Papers Given at a Review Symposium on Colloid Science, Organised by the Continuing Education Committee of the Royal Society of Chemistry

University of Bristol, 8th—10th September 1981

Edited by
J. W. Goodwin
University of Bristol

The Royal Society of Chemistry
Burlington House, London W1V 0BN

British Library Cataloguing in Publication Data

Colloidal dispersions.—(Special publication/Royal
  Society of Chemistry, ISSN 0260-6291; no. 43)
  1. Colloids—Congresses
  I. Goodwin, J. W.   II. Royal Society of Chemistry   III. Series
  541.3′451   QD549

ISBN 0-85186-865-7

Printed in Great Britain by Henry Ling Ltd., at the Dorset Press, Dorchester, Dorset

Preface

Most areas of scientific endeavour follow a pattern of
periods of rapid growth in ideas interspersed with longer periods
of consolidation. Colloid science has followed this rule with
several steps forward during this century. The 1940's were such
a period with the theories due to Derjaguin, Landau, Verwey and
Overbeek giving a sound basis for understanding the stability of
dilute suspensions which are stabilised by electrostatic forces
opposing the London - van der Waals' attractive forces. This
gave rise to a great deal of experimental work on a wide variety
of dilute dispersions verifying and consolidating these ideas.
The 1970's marks another important period of change for both
theoretical and experimental studies. For example, the ideas of
Lifshitz have become widely used to give a more comprehensive
description of attractive interparticle forces and there has been
a renewed interest in concentrated systems. Many of the ideas of
liquid state physics are currently being applied to concentrated
colloidal dispersions and this will be seen in the future to have
been a most profound development. In addition, new experimental
techniques such as dynamic light scattering and neutron scattering,
as well as a wider use of more conventional techniques such as
those found in rheology and particle counting are giving a
quantitative description of phenomena which can provide a thorough
test for the theories.

The last decade then has been a most exciting one in this
multidisciplinary field and it is a particularly apposite time
for a review of this field. As this volume is constituted of
the papers read at a "Review Symposium of Colloid Science" it is
of necessity limited in the number of contributions and some
topics are not represented here. However, the 'state of the art'
of most of the important topics is covered and we are particularly
fortunate in having such a distinguished list of contributors.

Professor Theo Overbeek has been responsible for many of the
advances in colloid science during the last fifty years.

Professors David Tabor and Douglas Everett have also had long and
distinguished careers in this field and have made important
contributions to the knowledge of attractive forces and adsorption
phenomena respectively. The other contributors, Professors
Hans Lyklema, Don Napper, Ron Ottewill and Dr. Peter Pusey are
also well known to the student of colloid science for their
important contributions. The symposium, therefore, was a most
interesting one and this is reflected in this volume.

Bristol 1981   J.W. Goodwin
                                                   Editor

Contents

1

Colloids, A Fascinating Subject: Introductory Lecture

By J. Th. G. Overbeek

VAN'T HOFF LABORATORY, PADUALAAN 8, POSTBUS 80051, 3508 TB, UTRECHT, THE NETHERLANDS

Summary

Colloid science may be considered to start with the study of *pseudo-solutions* in water of sulphur, silver chloride, Prussian blue (SELMI, 1845), of colloidal gold (FARADAY, 1857) and with the coining of the term "colloid" by GRAHAM in 1861. Colloids are present as dispersed systems, characterized by slow diffusion and slow (often negligible) sedimentation under normal gravity, which set the size of the colloidal particles in the range of about 1 nm to 1 μm.

Some colloid systems such as polymer solutions and soap solutions, containing *micelles*, form spontaneously. They are thermodynamically stable, and are called *lyophilic colloids*. Others, which contain particles, insoluble in the solvent (e.g. AgCl, S, Au, oil in water), must be prepared in a roundabout way. These are called *lyophobic colloids*. In their preparation the presence of *peptizing* or *stabilizing* substances is essential. Since van der Waals forces always tend to lead to agglomeration (flocculation) of the particles, stability (i.e. a long shelf life) of such colloids requires that the particles repel one another, either by carrying a net electrostatic charge or by being coated with a sufficiently thick adsorbed layer of large molecules compatible with the solvent.

Potential energy curves are very useful for the display of the various types of interaction between colloidal particles.

Colloidal dispersions have interesting optical properties, such as TYNDALL light scattering, turbidity and birefringence. They often show remarkable non-Newtonian rheological behaviour.

In the many practical applications in paints, inks, emulsions, pharmaceutical, agricultural and cosmetic preparations, photographic films, foams, etc. the stability and the rheological properties are of the utmost importance.

Colloidal aspects are numerous in biology, biopolymers, biological membranes and cells in general.

1.   Introduction

Colloid Science finds its origin in the discovery of solutions of
certain insoluble substances. SELMI[1] described in 1845 *pseudosolutions*
in water of silver chloride, sulphur and Prussian blue. He included them
in the same class as solutions of albumin and starch. All these solutions
are clear or slightly turbid. SELMI considered them to contain particles
which are larger than the usual molecules. MICHAEL FARADAY[2], towards the
end of his life, made fairly extensive studies of the colloidal gold sol
and to the present day the very fine gold sols (particles of about 3 nm
radius) prepared by reduction of a gold chloride solution with phosphorus
are called *Faraday sols*. GRAHAM[3] who coined the term "colloid" (= glue
like) in 1861 emphasized the low rate of diffusion and the lack of cry-
stallinity of all these colloidal systems and from this slow diffusion
drew the correct conclusion that colloid particles are fairly large (say
larger than 1 nm). On the other hand the fact that these same colloidal
solutions do not sediment (or only very slowly) under normal gravity
proves that the particles cannot be very large (say not larger than 1 μm).

Colloidal suspensions ($\equiv$ sols) have found numerous fundamental applications, as in PERRIN's studies[4] of Avogadro's constant, where colloidal particles acted as very large, individually visible molecules, or,
very recently, in VRIJ's use of concentrated colloidal suspensions as
models for the liquid state[5]. The rate of rapid coagulation of colloids,
in which each Brownian encounter between two particles results in permanent contact, was clarified in 1916 by the beautiful theoretical work
by VON SMOLUCHOWSKI[6]. This rate of coagulation is essentially identical
to the rate of diffusion - controlled bimolecular reactions between small
molecules[7]. A small but necessary refinement to the theory, which takes
into account that the last approach between two particles or molecules
is slowed down, because it is difficult to squeeze out the last layers
of liquid from between the particles, was recently formulated by several
authors (DERYAGIN and MÜLLER[8], SPIELMAN[9], HONIG, ROEBERSEN and WIERSEMA[10],
DEUTCH and FELDERHOF[11]) and nicely confirmed by experiments on the coagulation of latex particles (LICHTENBELT et al.[12]). Phase transitions
(HACHISU[13]) between concentrated liquid-like colloidal dispersions and
crystalline arrangements of the particles have been observed and they
follow from the statistical thermodynamics of such systems[14].

Other applications of colloids are eminently practical. Emulsions
(dispersions of liquid particles in a liquid, such as oil in water or
water in oil) are used very widely. Milk, mayonnaise, crude oil containing water at the well head, and many cosmetic and pharmaceutical
preparations belong in this category. Latices of various polymers are

prepared in quantities of millions of tons per year in a process called emulsion polymerization, in which one starts with an emulsion of the monomer in water.

Suspensions (dispersions of solid in liquid) are found in the kitchen in the form of various soups and sauces. Many paints, inks, and lacquers are fine suspensions of the pigments in oil or water or even in an emulsion. Clay particles are of colloidal size. It is hardly possible to exaggerate their importance in agriculture, in ceramics and even in delta formation.

Fogs and smokes are colloidal dispersions in gases. They are called *aerosols*.

Soap bubbles, foams and biological membranes show colloidal properties, since the thickness of the lamellae ranges from 4 nm to over 1 μm.

Colloidal systems have interesting optical properties, colour, turbidity, birefringence. Some of these properties are peculiar because the wavelength of visible light lies in the colloidal size range. Many colloidal particles carry an electric charge due to surface dissociation or to the adsorption of ions. This electric charge happens to be very important in stabilizing colloidal dispersions, in characterizing the particles and in several applications.

Colloids, especially concentrated colloids, have remarkable mechanical properties. Think of gelatin solutions, which flow freely above 40°C, but become rigid (form a *gel*) at room temperature. A good paint must flow easily when applied, but then must rigidify soon in order to prevent it from flowing off vertical surfaces. Wet clay must not change its shape in normal gravity, but yield easily to the sculptor's thumb.

I hope that this introduction has shown that the study of colloids is worthwhile to increase our understanding of nature and to allow us to control it more efficiently and that it has whet your appetite to taste more of this field of neglected dimensions. C.f. WO. OSTWALDs book, *Die Welt der vernachlässichten Dimensionen*[15].

2. Types of colloids and their preparation

The simplest type of colloid is formed by solutions of *macromolecules* (polymers), including biopolymers such as proteins, nucleic acids and polysaccharides. These systems belong to colloid science since the sizes of macromolecules fall in the colloidal size range. The older generation of colloid scientists (FREUNDLICH,[16], KRUYT and BUNGENBERG DE JONG[17]) considered them as an important part of their science, but more recently on account of the great technical importance especially of man-made polymers, polymer science has developed as a science on its own with its own textbooks (e.g. FLORY[18]) and periodicals.

However, colloid science is not complete without the inclusion of polymers, not only because techniques such as light scattering, sedimentation, and electrophoresis are applied to polymers just as to many other colloids, but also because of the many other links. There are polymer colloids (FITCH[19]), in which particles of colloid size made of polymers are dispersed in non-solvents for the polymer molecules. Furthermore polymers are frequently used to modify the properties of other colloids as we shall see in section 3 on steric stabilization.

The next class of colloids consists of solutions of soaps and other *amphipolar* substances. These molecules are characterized by a large non-polar part (hydrocarbon or fluorocarbon) and a water compatible polar part. The polar part may be anionic, as in $C_{12}H_{25}OSO_3Na$, or cationic, as in $C_{16}H_{33}N(CH_3)_3Br$, or non-ionic as in $C_9H_{19}(C_6H_4)O(CH_2CH_2O)_8H$. In water, above a certain concentration (the lower, the larger the non-polar part), aggregates of 20 or more molecules, which are called *micelles*, are formed. Therefore these colloids are called *association colloids*. The

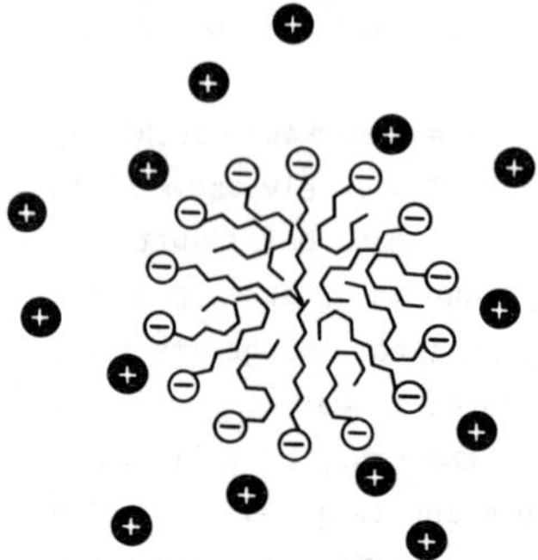

Figure 1. Schematic representation of a spherical or cylindrical micelle.

lack of solubility of the hydrocarbon (fluorocarbon) part in water drives the non-polar groups together, but instead of then forming a separate phase the particle growth stops when the non-polar nucleus is sufficiently surrounded by polar groups, thus minimizing the water-hydrocarbon contact area. The same substances may form *inverse micelles* (polar parts inside) in non-polar solvents.

Micelles and inverse micelles may take up other molecules in their interior in a process called *solubilization*. This may even go so far as to result in the formation of fairly large droplets with a radius of say 10 nm. Such systems are called *microemulsions* and may be either of the water-in-oil or oil-in-water type. In recent years microemulsions have drawn a great deal of attention, since they are applied in tertiary oil recovery on the basis of the extremely low interfacial tensions (to less than 0.001 mN m^{-1}) exhibited by these systems.

All the categories mentioned so far are thermodynamically stable colloids and are known as *lyophilic colloids* (*hydrophilic* if water is the dispersion medium).

Colloidal dispersions of "insoluble"substances, called *lyophobic* (*hydrophobic*) *colloids*, are not in thermodynamic equilibrium, but, if well prepared, they may have life times of many years. (Some of FARADAY's gold sols still exist). Lyophobic colloids are prepared either by *dispersion methods* (= grinding, milling, etc. until the particles are small enough) or - more subtly - by *condensation methods*. In a condensation method the insoluble material is precipitated from a solution of small molecules or ions under circumstances in which a high rate of nucleation of the new phase is combined with a relatively slow rate of growth of these nuclei. The process may sometimes be carried out in such a way that all the nuclei are formed in a very short time early in the process, whereas the growth to larger particles occurs over a longer time without further nucleation. Under these circumstances all particles grow at the same rate which leads to *monodisperse* (= *isodisperse*) systems. The method goes back to ZSIGMONDY's[20] gold sols, but it was applied to many other cases (sulphur sols, aerosols) and interpreted more fully by LA MER and coworkers[21]. Figure 2, based on LA MER's work[22] shows that, if the material for the particles is supplied continuously (e.g. by slow formation of sulphur from acidified thiosulphate) and if the nucleation concentration is sufficiently far above the saturation concentration, a short burst of nucleation may rapidly lower the concentration below the supersaturation needed for nucleation, after which all further material will only cause growth of the existing particles, without out any new ones being formed.

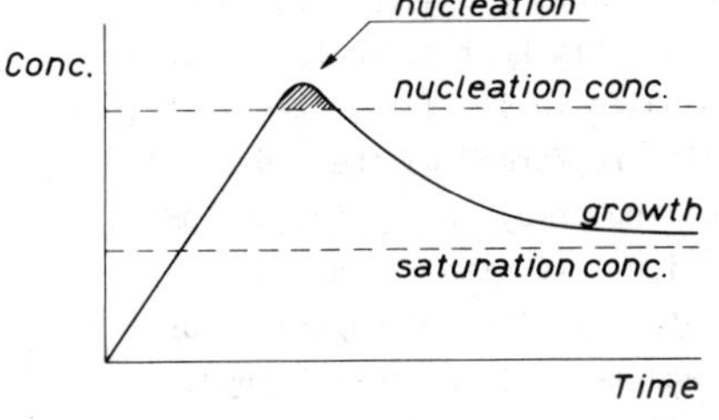

Figure 2. Showing schematically how a nucleation period and a growth period may be separated, thus leading to a monodisperse sol.

Whether prepared by dispersion or by condensation lyophobic colloids only have a long life if suitable *peptizing* or *stabilizing substances* are present. Such substances are needed to prevent the particles from making actual contact and sticking together during their frequent Brownian encounters. The vivid Brownian motion of colloidal particles forms a fascinating sight, when observed in a dark field microscope (an *ultramicroscope*) and has been described as a tableau of "thousands of dancing gnats". Figure 3 shows the central part of the field of view in SIEDENTOPF and ZSIGMONDY's ultramicroscope[23], in which the beam of a strong light source is concentrated by a microscope objective perpendicular to the line of sight of the microscope through which the phenomenon is observed.

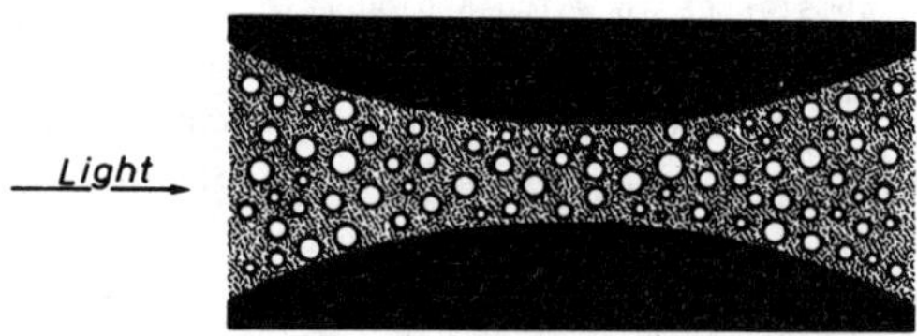

Figure 3. Central part of the field of view in SIEDENTOPF and ZSIGMONDY's ultramicroscope.

We now give a few examples of the preparation of lyophobic colloids.

Oil-in-water (O/W) and water-in-oil (W/O) *emulsions* are prepared by shaking, stirring, or sonication of a mixture of oil and water. The resulting emulsion is only stable if some soap, or some other stabilizer, is present to prevent the newly formed oil-water interfaces from coalescing. The type of stabilizer determines whether an O/W or W/O emulsion is formed.

Silver halide sols are formed by mixing solutions of silver nitrate and alkali halide. They require an excess of Ag^+ or Hal^- ions for their stability.

The *arsenic sulphide sol* is prepared by pouring a saturated solution of As_2O_3 into a solution of H_2S in water which is kept saturated by bubbling H_2S through it. The excess H_2S (sulphide ions) is the stabilizer.

Ferric hydroxide and *ferric oxide* sols are formed by the hydrolysis of ferric chloride solutions either by heating or by increasing the pH. Depending upon the conditions Fe^{3+} or OH^- ions are stabilizing ions.

Gold sols are formed by reducing dilute gold (Au^{III}) chloride solutions with a great variety of reducing agents, such as formaldehyde, hydrazine, hydrogen peroxide, sodium citrate. The stabilizer is excess chloride or a gold chloride or gold citrate complex ion.

Polymer latices are usually prepared by free radical polymerization of vinyl and/or butadiene monomers in an emulsion or solution containing soap or soap-like substances, which act as stabilizers for the polymer particles (radius e.g. 100 nm) formed.

From these examples it may appear that the stabilizing substances provide ions that are adsorbed onto the particles, thus charging them, so that they repel one another. There are, however, other very general stabilizing substances, such as gums and gelatin which are non-specific *protective agents* for dispersions in water and which may protect even without being electrically charged.

This brings us to suspensions in oil and in other *non-aqueous media*, which are usually prepared by grinding in the presence of suitable, oil-soluble and adsorbable protective agents. Examples are paints and inks, doped engine oils in which the dopant, a protective agent, keeps carbon particles resulting from incomplete combustion in suspension, thus preventing them from causing abrasion.

Since many colloids contain contaminating substances that have remained from the preparation stage, it is desirable and sometimes essential to remove these substances. This can be accomplished by *dialysis*, i.e. by bringing the colloid in contact with pure, frequently renewed, solvent via a membrane with pores, small enough to retain the colloid particles, but large enough to let small molecules or ions pass through. The dialysis can often be accelerated by applying an electric field across the membrane. Figure 4 shows GRAHAM's original dialyzer[24], an inverted bottomless beaker closed at one side by a parchment (later cellophane or collodion) membrane. Figure 5 shows a modern five chamber electrodialyzer[25].

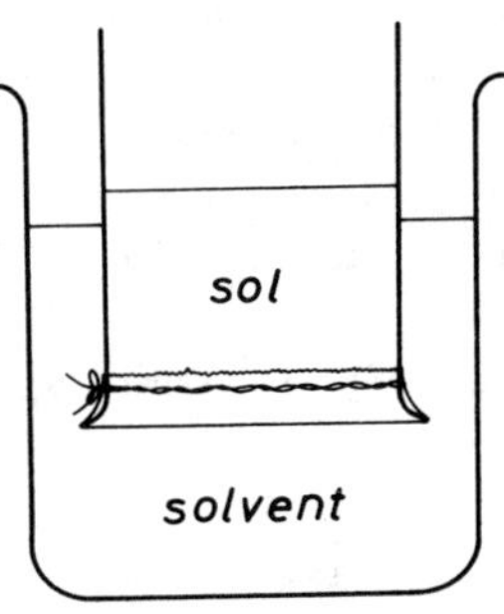

Figure 4.   Dialyzer as used by GRAHAM.

In a very simple, but effective method of dialysis the sol is enclosed in a length of regenerated cellulose or nitrocellulose tubing ("artificial

sausage skin"), which is allowed to float around in the frequently renewed solvent.

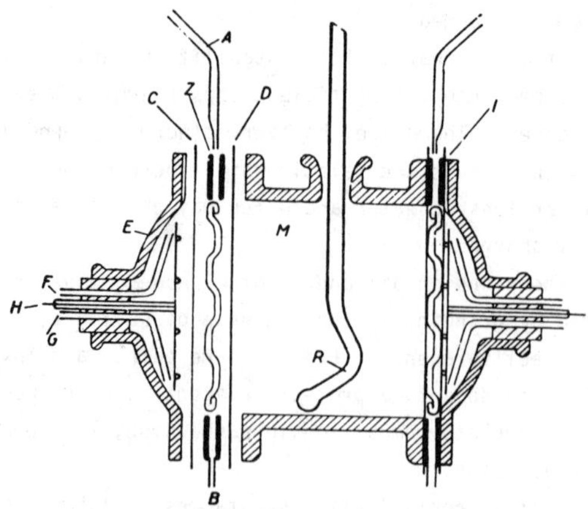

Figure 5. Electrodialyzer with five chambers.
A,F = supply of distilled water, B = drain of rinsing chamber,
C,D = membranes, E = electrode chamber, G = drain of electrode
compartments, H = electrode, I = groove for escape of gas, M = middle
chamber, containing sol, R = stirrer.

3. Stability of lyophobic colloids

Colloid particles always attract one another by *van der Waals forces*
(*dispersion forces*), even across a condensed medium, although the medium
weakens the force. Due to the near additivity of the forces between pairs
of molecules, the attractive force has a fairly long range. Between two
spherical particles of radius a at a distance R between the centers and a
distance $H = R - 2a$ between the surfaces, the energy of attraction is
still of the order of kT (k = the Boltzmann constant, T = the absolute
temperature) when $H \simeq 0.1a$.

HAMAKER derived the following expression for the energy of attraction
between two equal spherical particles[26].

$$\Delta V_{att} = - \frac{A}{6} \left[\frac{2a^2}{R^2-4a^2} + \frac{2a^2}{R^2} + \ln \frac{R^2-4a^2}{R^2} \right] \tag{1}$$

where the *Hamaker constant*, A, depends strongly on the nature of the
particles and the medium, but is usually in the range of 10^{-21} -10^{-19}J,
or $A \sim 0.25$ kT to 25 kT at room temperature.

For small distances eq.(1) can be approximated by

$$\Delta V_{att} \simeq - \frac{Aa}{12H} \qquad (2)$$

or more precisely[27]

$$\Delta V_{att} \simeq - \frac{A}{12} [\frac{L}{H} + 2 \ln \frac{H}{L}] \qquad (3)$$

where $L = a + 3H/4$.

Given the intensity and the range of the van der Waals attraction, a sol can only be stable if the particles also repel one another with a force of sufficient strength and sufficiently long range.

Two types of repulsion are known and relatively well understood, an electrostatic and a steric repulsion. The *electrostatic repulsion* is based on the charge of the particles, obtained by surface dissociation or by preferential adsorption of ions of one type. It is, however, not a simple Coulomb repulsion, since, to conserve electroneutrality, the particle charge is surrounded by a more or less diffuse ion atmosphere, which forms an *electric double* layer with the surface charge. The double layer is represented in figure 6. It may be divided into a molecular

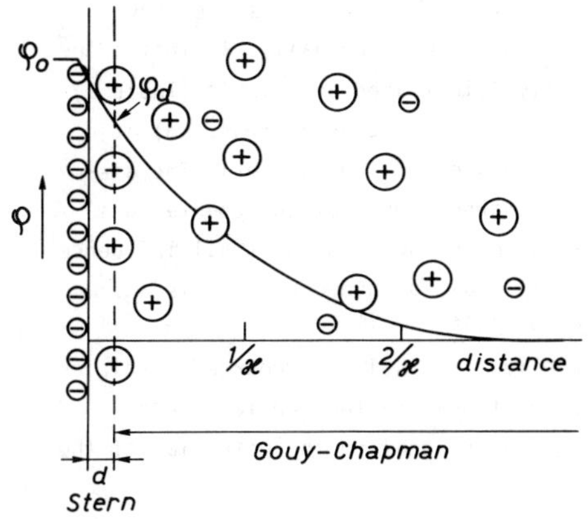

Figure 6. Schematic representation of the distribution of the ions and the decay of the electric potential ϕ in a double layer.

condenser or STERN layer[28] and a diffuse or GOUY-CHAPMAN layer[29]. At large distances, x, from the surface the potential decays as $\exp(-\kappa x)$, at short distances somewhat more steeply. κ is the inverse DEBYE-HÜCKEL length[30]. It is given by

$$\kappa = \left[\frac{F^2 \Sigma \, c_i z_i^2}{\varepsilon_r \varepsilon_o \, RT} \right]^{\frac{1}{2}} \tag{4}$$

in which c_i and z_i are the concentration and charge number resp. of the ions of type i, ε_r the dielectric constant of the solution, ε_o the permittivity of the vacuum, R the gas constant and F the Faraday constant.

The energy of repulsion between two particles carrying double layers decays approximately as $\exp(-\kappa H)$ and can be written[31] *)

$$\Delta V_{rep} \simeq + 2\pi \, \varepsilon_r \, \varepsilon_o \, a \, \left(\frac{4RT}{zF} \, \gamma \right)^2 \exp(-\kappa H) \tag{5}$$

where γ = tanh $(zF \, \phi_d / 4RT)$, z is the charge number of the counter ions and ϕ_d the potential at the plane of the Stern-ions (dissolved counter ions in closest approach to the surface, see fig. 6).

Because the attraction decays as an inverse power of the distance (as H^{-1} for spherical particles at short distances, and as H^{-6} at large distances) but the repulsion decays exponentially with H, the attraction prevails both at short and at large distances, but at intermediate distances $(H \simeq \kappa^{-1})$ the repulsion may win and an energy barrier is formed which keeps the particles apart and thus stabilizes the sol. Increasing the concentration and/or the charge number of the electrolyte in the solution *compresses the double layer* $(\kappa \simeq z \sqrt{c}\,)$, shortens the range of the repulsion, tends to decrease ϕ_d, and thus eventually eliminates the energy barrier. This is most strikingly illustrated by a table (Table 1) of electrolyte concentrations (*critical coagulation concentrations*) at which a sol loses its stability and coagulates (= flocculates). The disproportionately large influence of the charge number of the counterions (= ions charged oppositely to the particle charge) and the small influence of the specific nature of the ions is typical for hydrophobic sols and is known as the *rule of Schulze*[35] *and Hardy*[36].

When a d.c. electric field is applied to a sol with charged particles, the particles move with a velocity proportional to the applied field strength and in a first approximation proportional to the potential of the

---

*   This equation is an approximation in several respects. It assumes that ϕ_d is independent of H, whereas $|\phi_d|$ in general will increase with decreasing H. Moreover even at ϕ_d = constant the repulsion would increase more steeply than $\exp(-\kappa H)$ for small values of H. However eq.(5) is a good approximation at large κH and it predicts the effects correctly in a semi-quantitative sense at smaller κH.

Table 1

Critical coagulation concentrations (c.c.c.) in millimol/l.

As_2S_3 - sol, negatively charged[32]

	c.c.c.		c.c.c.		c.c.c.
LiCl	58	$MgCl_2$	0.72	$AlCl_3$	0.093
NaCl	51	$CaCl_2$	0.65	$\frac{1}{2}Al_2(SO_4)_3$	0.096
KNO_3	50	$ZnCl_2$	0.69	$Ce(NO_3)_3$	0.080

Au - sol, negatively charged[33]

	c.c.c.		c.c.c.		c.c.c.
NaCl	24	$CaCl_2$	0.41	$\frac{1}{2}Al_2(SO_4)_3$	0.009
KNO_3	25	$BaCl_2$	0.35	$Ce(NO_3)_3$	0.003
$\frac{1}{2}K_2SO_4$	23				

Fe_2O_3 - sol, positively charged[34]

	c.c.c.		c.c.c.
NaCl	9.25	K_2SO_4	0.205
KCl	9.0	$MgSO_4$	0.22
KBr	12.5	$K_2Cr_2O_7$	0.195

mobile part of the double layer, the so-called zeta(ζ) potential. The
ζ - potential is usually close to the Stern - potential, ϕ_d. This motion
is called *electrophoresis*. It is one of the *electrokinetic phenomena*.
They all have their origin in the double layer and manifest themselves
when either an electric field or a flow field is applied parallel to the
interface. Electrophoresis is often a valuable source of information on
the sign and the value of the charge and the potential of the double
layer.

We now consider the other type of repulsion, *steric repulsion*. It is
obtained when the particle surface is covered (by adsorption or chemical
reaction) by bulky molecules, usually long chain molecules, thus giving
the particles a "hairy" surface. If these hairs are compatible with
(= soluble in) the medium, they repel one another and they cause a steep
repulsion between the particles, with a range of the order of the size
of the randomly coiled hairs. As sketched in figure 7 the repulsion is
due to two effects, an *osmotic effect* caused by the high concentration of
chain elements in the region of overlap, and a *volume restriction effect*
due to the loss of possible conformations in the narrow space between the
two surfaces [37,38].

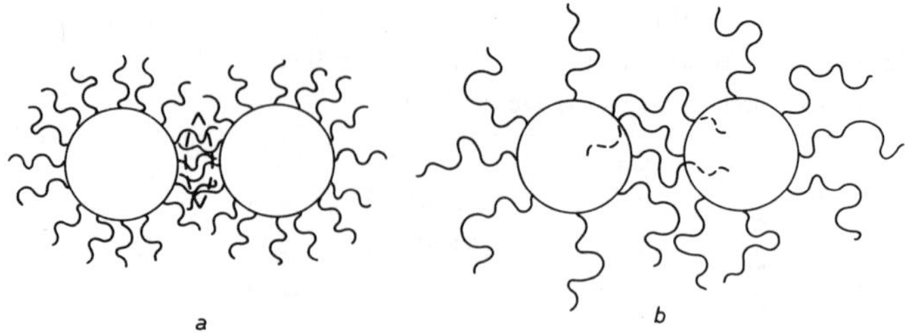

Figure 7. Illustrating two aspects of steric repulsion. *a.* Osmotic
effect due to high concentration of chain elements in the region of overlap. *b.* Volume restriction effect due to loss of possible conformations
of the "hairs".

Steric repulsion is particularly important for stabilization in nonaqueous, non-polar media. It can be influenced by changing the length
and nature of the hairs, by influencing the adsorption energy, and by
changing the properties of the solvent. In worse than thêta conditions
the hairs collapse and steric repulsion is not effective. Substances
causing steric stabilization are either of the *anchor and chain* type or
homopolymers. The anchor is easily adsorbed, the chain soluble in the
medium (soaps are examples, so are suitable block copolymers). With the
homopolymers a fraction of the chain elements is adsorbed (as *trains*),
most of the elements are in the solvent as *loops* and *tails*. Protective
agents mentioned earlier act by steric repulsion, even in water.

In the last few years renewed interest[39] has arisen in the so called
structural forces. These are due to the modification of the solvent
structure near interfaces as a result of packing restrictions. As far
as can be judged now, these forces in themselves probably do not lead to
a repulsion of a sufficiently long range for stability, but they may have
a comparatively large effect at small separations between the particles
where they have an oscillating character and they may play an essential
role in *repeptisation* (= redispersion of a coagulated suspension or
emulsion).

4. <u>Potential energy curves</u>

Potential energy curves, in which the change in free energy (ΔV)
is plotted against the distance (H) between the particle surfaces, form a
powerful instrument for surveying the interplay between different forces.

Figure 8 shows a set of such curves for the repulsion, the attraction
and the combination of the two in the case of electrostatic stabilization.
If the particles are caught in the very deep minimum (the *primary mini-*
mum) at small separations ($\Delta V_{total} \rightarrow -\infty$ as $H \rightarrow 0$) it is very difficult

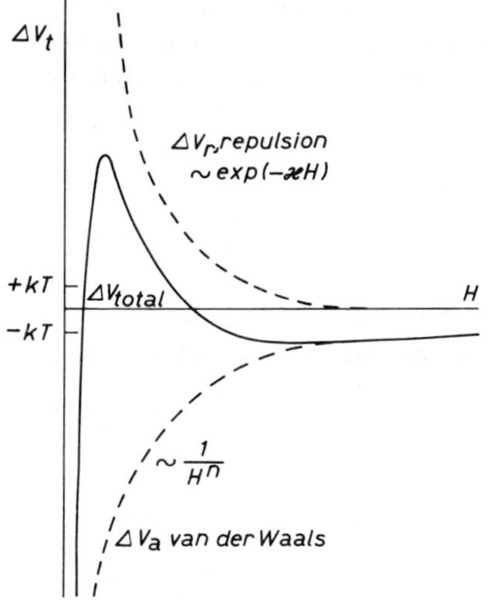

Figure 8. Electro-
static repulsion, van
der Waals attraction
and the combination
$\Delta V_{total} = \Delta V_r + \Delta V_a$.

or impossible to separate them. The shallow minimum at large H (the
secondary minimum), which may be only a fraction of kT deep, acts as a
temporary trap, from which the particles can be removed easily by mechanical means, change of composition of the medium, or even simply by Brownian
motion. If the maximum in the curve is more than 15 or 20 kT, effective
collisions leading to permanent contact in the primary minimum become so
rare that the suspension is stable. The effect of changes in the parameters of the system (electrolyte composition, Hamaker constant) on the
potential energy curves can be followed easily and then conclusions can
be drawn on the behaviour of the actual system.

Figure 9 shows the corresponding curves for steric stabilization.
The repulsion is very steep, the primary minimum often unattainable.

Figure 10 is an example of the possible form of the free energy curve
for structural forces. The wavelength of the oscillations corresponds to
the distance between the molecules in the solvent in the neighbourhood of
an interface. Depending on how difficult or easy it is to remove the last
layer of molecules between two interfaces the structural force may be repulsive or attractive at very small separations. The subject is in rapid

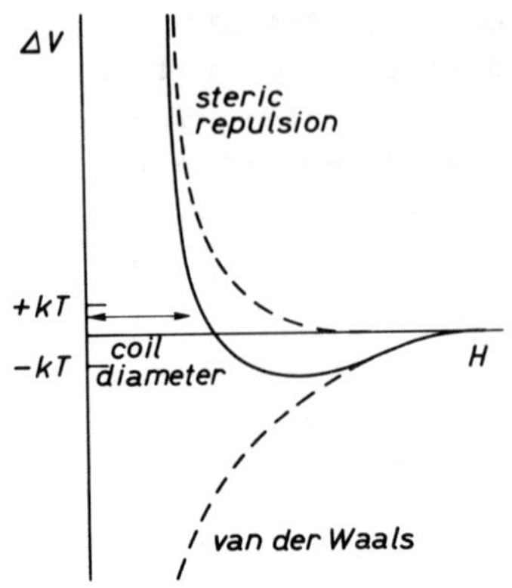

Figure 9. Schematic representation of the free energy as a function of H in the case of steric stabilization.

development, both theoretically and experimentally, but has not quite advanced to the point for application to the stability of real examples, say that of silica particles in water.

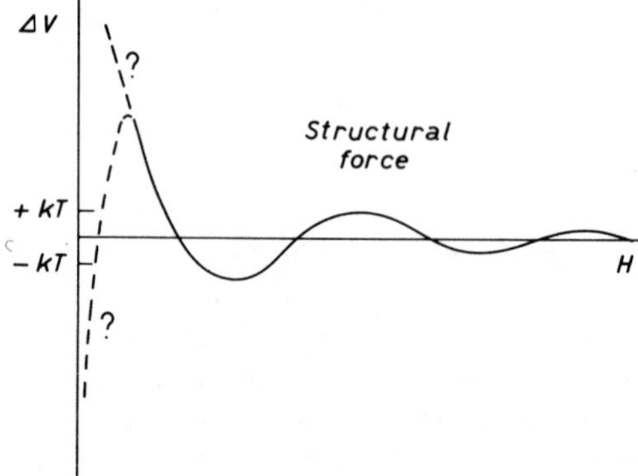

Figure 10. Schematic representation of the free energy curve for structural forces.

5.   <u>Properties of colloids and some applications</u>

Most colloid suspensions, emulsions and aerosols are *turbid*. They scatter light (TYNDALL light[40]) in all directions. For small particles the scattered intensity is proportional to the square of the volume of the particles. The scattering per unit volume of the suspension is thus proportional to its mass concentration times the volume (or mass, or *mole-*

cular weight) of the particles. For larger particles, where particle
size and wavelength of light are of comparable dimensions, the total
intensity and the angular dependence of the scattered light become very
complicated (MIE theory[41]). The turbidity of colloids is applied in a
variety of ways. It is used for the determination of the molecular weight
and, if large enough, the size of the particles. Fibers such as nylon or
polyester are made opaque by the incorporation of finely divided TiO_2.
The back scatter of the head lamp is known to everyone who has driven a
car at night in a fog. Aerosols are used for military camouflage purposes,
but also for the protection of orchards against frost damage by excessive
radiation to the clear night sky.

The availability of lasers has made it possible to determine the
diffusion coefficient of colloidal particles by analyzing the shift in
wavelength of the scattered light caused by the Doppler effect or by
photon correlation spectroscopy[42].

Colloids are often deeply coloured (red gold sol, deep yellow As_2S_3
sol, red V_2O_5 sol) if the colloid material absorbs visible light. Paints
and inks in which the pigment particles are finely dispersed are an obvious application of this effect.

As mentioned before colloids show *electrophoresis* in an electric
field. In such a field the double layer is polarized as shown in figure
11, but not permanently stripped off the particle. This polarization

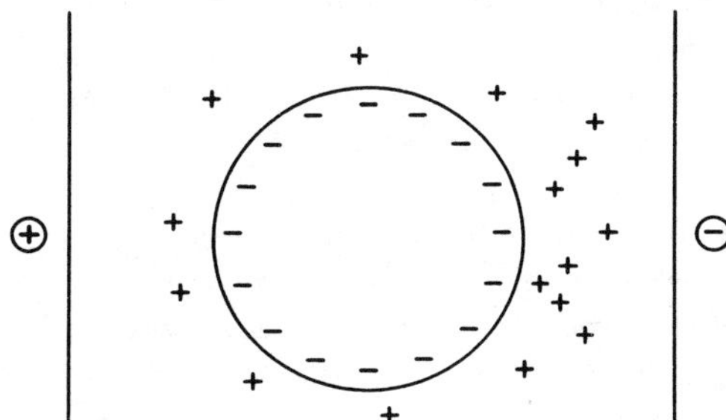

Figure 11. Showing
how the double
layer around a
particle is polar-
ized in an electric
field.

affects the electrophoretic mobility of the particle. The dipoles formed
cause an increase (in some cases a very large increase) in the dielectric
constant of the system. If the particles are rod-like or plate-like the
polarization tends to orient the particle parallel to the lines of force

(see figure 12) thus causing electric birefringence (Kerr-effect[43]).

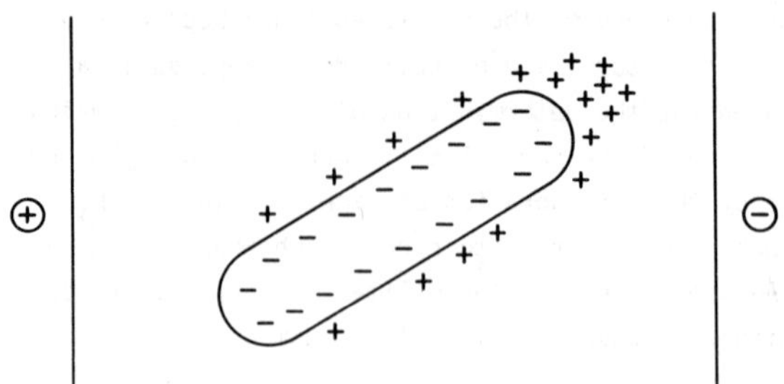

Figure 12. The electric polarization of an anisodimensional particle leads to orientation parallel to the field and thus to electric birefringence (Kerr effect).

Electrophoresis finds analytical and preparative applications in sorting out the complicated mixtures of proteins and other substances in biochemistry. In *electrodeposition* paint suspensions are deposited by electrophoresis onto metal substrates. The process is used on a large scale in painting car platework.

Suspensions of *ferromagnetic* particles become birefringent by orientation of the particles in a magnetic field (Cotton-Mouton effect). Concentrated suspensions of this nature may flow freely in zero field but become rigid in a strong magnetic field.

Correct preparation and handling of suspensions of magnetic particles is essential for the preparation of sound tapes and videotapes.

Concentrated colloid systems have remarkable *rheological properties*. Stable suspensions and emulsions behave as Newtonian liquids in weak shearing fields, but under strong shear the particles cannot move fast enough past one another and the system rigidifies (shear thickening or *dilatancy*). If one walks on a wet sandy beach one sees a ring of dry sand around each footstep caused by dilatancy of the wet sand.

Stable systems settle eventually to a very *dense sediment* since the particles can roll over each other until the closest packing is reached. This may happen in a can of paint which is too well stabilized. If the system is unstable, even when it is only "weakly flocculated", the sediment volume is large and it may even fill the whole suspension, thus causing the formation of a *gel* as illustrated in figure 13. If such a weakly flocculated suspension is stirred or shaken it may become fluid because

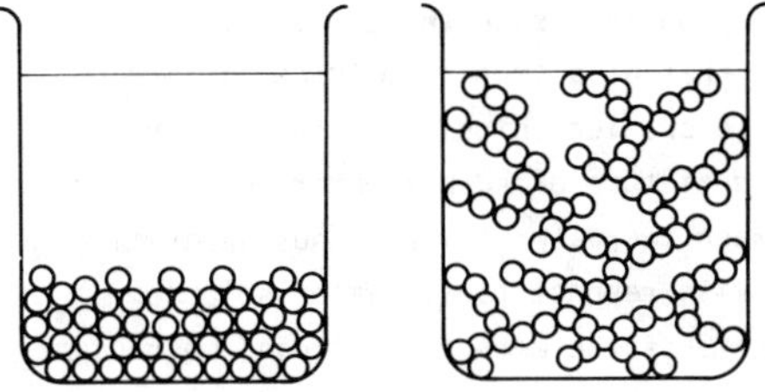

Figure 13. A stable suspension forms a dense sediment, a flocculated suspension has a large sediment volume and may even form a gel.

a sufficient number of bonds between particles is broken. After being left undisturbed for some time it returns to the gel state. This is called *thixotropy*[44] or *reversible sol-gel transformation*, a property essential for a good paint, for the drilling muds used in oil well drilling, and in many pastes.

Soap bubbles and foam lamellae would break very quickly, if they were not stabilized by the adsorption of the amphipolar soap or soap-like molecules at their surfaces. Figure 14 illustrates for an ionized soap how the overlap of the two double layers is similar to that between electrically stabilized colloidal particles and how this may prevent the ultimate thinning and breaking of the lamella.

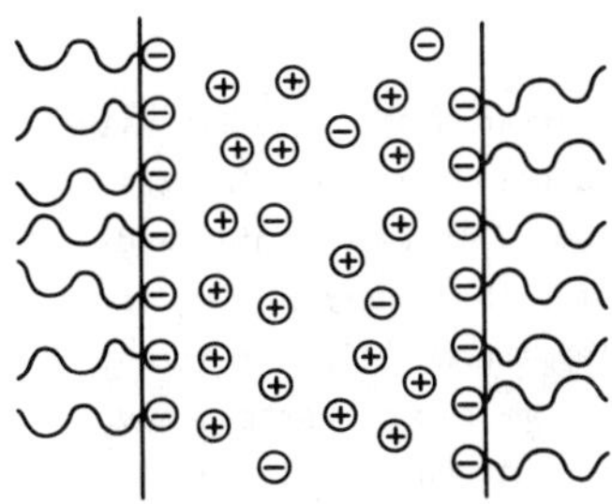

Figure 14. Cross section through a soap film stabilized by the adsorption of soap ions at its surfaces and by the overlap of double layers.

Foams are sometimes a nuisance and must be destroyed or their formation prevented as in distillations or in washing machines. They also find many useful applications as in *froth flotation*, a process by which yearly millions of tons of minerals are separated from gangue and from one another, in beer, whipped cream and many other applications in foods, also in fire fighting foams.

Foams and soap films bring us close to at least one application of colloids in biology. All *biological cells* are surrounded by a membrane and similar membranes occur copiously inside the cell. The essential structure of those membranes is that of a soap film turned inside out. The amphipolar molecules in this case are lecithins and other phospholipids. Figure 15 gives a schematic picture of such a membrane. As can be seen it forms an oily layer between two water layers, thus separating the inside and outside of the cell with respect to all water soluble molecules. If such molecules have to pass the membrane, special mechanisms are locally built into the membrane.

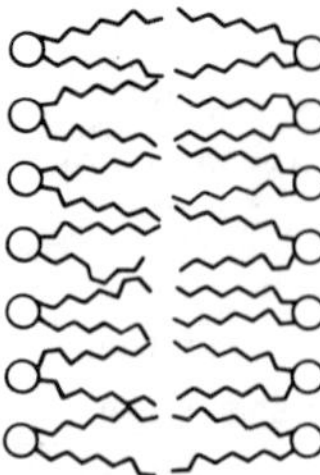

Figure 15. Biological membrane built from phospholipids with two hydrocarbon chains and a polar part which is zwitterionic (phosphate and ammonium) in lecithin.

Other colloidal aspects in biology are the biopolymers (proteins, nucleic acids, polysaccharides), cell adhesion, rheology of blood, the emulsion aspect of milk etc.

6. Conclusion

The description of colloid systems as dispersions of particles with interparticle forces of various origins forms a good framework for the understanding of many properties of these systems. Thermodynamics and statistical thermodynamics are important both for the understanding of lyophilic colloidal solutions as a whole and for the detailed treatment of interparticle forces in lyophobic systems. A number of aspects have been treated incompletely or not at all in this introductory lecture. They will be treated more fully in later lectures in this symposium.

Colloids find a great many applications, some based on the small size and the large number of the particles, some on the huge interfacial area with its capacity for heterogeneous reactions, others again on their optical, electrical or mechanical properties. Not a few of these applications are found in fundamental research.

Colloids are fascinating to look at, fascinating to understand, fascinating in their applications.

Acknowledgements

I want to express my gratitude to Prof. P.L. de Bruyn for his stylistic and technical criticisms, to Mr. N.M. van Galen and to Mr. M. Tollenaar for making the drawings and the slides, to Mr. J. Suurmond for preparing the demonstrations and to Miss Shana Abrams for the great care with which she has prepared the typescript.

References

[1] F. Selmi, Nuovi Ann. Science Natur. di Bologna (2), 1845, 4, 146; 1847, 8, 404; Ann. di Majocchi, 1844, 15, 88, 235; 1846, 24, 225. See also: E. Guareschi, Kolloid-Z., 1911, 8 113.

[2] M. Faraday, Phil. Trans. Royal Soc., 1857, 147, 145.

[3] Th. Graham, Phil. Trans. Royal Soc., 1861, 151, 183.

[4] J. Perrin, Ann. de Chim. et de Phys. (8), 1909, 18, 5.
J. Perrin, "Les Atomes", 9 me mille, Libr. Félix Alcan, Paris, 1920.

[5] A.A. Caljé, W.G.M. Agterof and A. Vrij, in "Micellization, Solubilization and Microemulsions", K.L. Mittal, ed., Plenum Press, New York, 1977, Vol. 2, p. 779.
A. Vrij, E.A. Nieuwenhuis, H.M. Fijnaut and W.G.M. Agterof, Faraday Disc. of the Chem. Soc., 1978, 65, 101.

[6] M. von Smoluchowski, Physik. Z., 1916, 17, 557, 585; Z. Physik. Chem., 1917, 92, 129.

[7] J. Halpern, J. Chem. Educ., 1968, 45, 373.

[8] B.V. Deryagin and V.M. Müller, Dokl. Akad. Nauk. S.S.S.R. (Engl. Transl.), 1967, 176, 738.

[9] L.A. Spielman, J. Colloid Interface Sci., 1970, 33, 562.

[10] E.P. Honig, G.J. Roebersen and P.H. Wiersema, J. Colloid Interface Sci., 1971, 36, 97.

[11] J.M. Deutch and B.U. Felderhof, J. Chem. Phys., 1973, 59, 1669.

[12] J.W.Th. Lichtenbelt, H.J.M.C. Ras, and P.H. Wiersema, J. Colloid Interface Sci., 1974, 46, 522; J.W.Th. Lichtenbelt, C. Pathmamanoharan, and P.H. Wiersema, 1974, 49, 281.

[13] S. Hachisu, Y. Kobayashi and A. Kose, J. Colloid Interface Sci., 1973, 42, 342; A. Kose and S. Hachisu, ibidem, 1976, 55, 487; S. Hachisu, A. Kose, Y. Kobayashi, and K. Takano, ibidem, 1976, 55, 499.

[14] J.G. Kirkwood, J. Chem. Phys., 1939, 7, 919;
B.J. Alder and F.E. Wainright, Phys. Rev., 1962, 127, 359;
B.J. Alder, W.G. Hoover, and D.A. Young, J. Chem. Phys., 1968, 49, 3688.

[15] Wo. Ostwald, "Die Welt der vernachlässigten Dimensionen", 8th ed.,
Steinkopff, Dresden, Leipzig, 1922; translated by M.H. Fischer as "An
Introduction to Theoretical and Applied Colloid Chemistry, The World of
Neglected Dimensions", 2nd ed. Wiley, New York, 1922.

[16] H. Freundlich, "Kapillarchemie", Band II, Akad. Verlagsges., Leipzig,
1932, p.3, 290.

[17] H.R. Kruyt, ed. "Colloid Science", Vol. 2, "Macromolecular and Association
Colloids", Elsevier, Amsterdam, 1949. H.G. Bungenberg de Jong, several
chapters in this Volume.

[18] P.J. Flory, "Principles of Polymer Chemistry", Cornell University Press,
Ithaca, 1953.

[19] R.M. Fitch, ed. "Polymer Colloids", Plenum Press, New York, Vol. 1, 1971,
Vol. 2, 1980.

[20] R. Zsigmondy and P.A. Thiessen, "Das kolloide Gold", Akad. Verlagsges.,
Leipzig, 1925.
R. Zsigmondy, Z. physik. Chem., 1906, 56, 65, 77.

[21] V.K. La Mer and M.D. Barnes, J. Colloid Sci., 1946 , 1, 71; V.K. La Mer,
J. Phys. Colloid Chem., 1948, 52, 65; D. Sinclair and V.K. La Mer, Chem.
Revs., 1949, 44, 245; V.K. La Mer and R. Gruen, Trans. Faraday Soc., 1952,
48, 410.

[22] V.K. La Mer and R.H. Dinegar, J. Amer. Chem. Soc., 1950, 72, 4847; Ind.
Eng. Chem., 1952, 44, 1270.

[23] H. Siedentopf and R. Zsigmondy, Ann. Physik (4), 1903, 10, 1. See also
R. Zsigmondy, "Zur Erkentniss der Kolloide," Jena, 1905.

[24] Th. Graham, see ref 3.

[25] H. de Bruyn and S.A. Troelstra. Kolloid-Z., 1938, 84, 192.

[26] H.C. Hamaker, Physica, 1937, 4, 1058.

[27] J.Th.G. Overbeek, J. Colloid Interface Sci., 1977, 58, 408.

[28] O. Stern, Z. Elektrochem., 1924, 30, 508.

[29] G. Gouy, J. phys., (4), 1910, 9, 457; Ann. phys. (9), 1917, 7, 129; D.L.
Chapman, Phil Mag. (6) 1913, 25, 475.

[30] P. Debye and E. Hückel, Physik. Z., 1923, 24, 185.

[31] E.J.W. Verwey and J.Th.G. Overbeek, "Theory of the stability of lyophobic
colloids", Elsevier, Amsterdam, 1948, p. 140. Essentialy the same equation
together with the subsequent treatment was given by B.V. Deryagin and L.
Landau, Acta Physicochim. U.R.S.S., 1941, 14, 633; J. Expt. Theor. Phys.
(in Russian), 1941, 11, 802, 1945, 15, 662.

[32] H. Freundlich, Z. physik. Chem., 1903, 44, 129; 1910, 73, 385.

[33] H. Freundlich and G. von Elissafoff, Z. physik. Chem., 1912, 79, 385; H.
Moravitz, Kolloidchem. Beihefte, 1910, 1, 301.

[34] H. Freundlich, Z. physik. Chem., 1903, 44, 151.

[35] H. Schulze, J. prakt. Chem. (2), 1882, 25, 431; 1883, 27, 320.

[36] W.B. Hardy, Proc. Roy. Soc. London, 1900, 66, 110; Z.physik. Chem., 1900, 33, 385.

[37] D.H. Napper, Trans. Faraday Soc., 1968, 64, 1701; J. Colloid Interface Sci., 1970, 32, 106; R. Evans and D.H. Napper, Kolloid Z.Z. Polym., 1973, 251, 329, 409.

[38] F.Th. Hesselink, J. Phys. Chem., 1969, 73, 3488; 1971, 75, 65; F.Th. Hesselink, A. Vrij and J.Th.G. Overbeek, J. Phys. Chem., 1971, 75, 2094.

[39] D.J. Mitchell, B.W. Ninham and B.A. Pailthorpe, Chem. Phys. Letts., 1977, 51, 257; J. Chem. Soc. Faraday Trans. II, 1978, 74, 1098, 1116; D.Y.C. Chan, D.J. Mitchell, B.W. Ninham and B.A. Pailthorpe, Mol. Phys., 1978, 35, 1669; J.Chem. Soc. Faraday Trans. II, 1979, 75, 556, 1980, 76, 776. W.J. van Megen and I.K. Snook, J. Chem. Soc. Faraday Trans. II, 1979, 75, 1095; I.K. Snook and W.J. van Megen, J. Chem. Phys., 1979, 70, 3099; 1980, 72, 2907; J. Chem. Soc. Faraday Trans. II, 1981, 77, 181; J.E. Lane and T.H. Spurling, Chem. Phys. Letts., 1979, 67, 107; Aust. J. Chem., 1980, 33.

[40] J. Tyndall, Phil. Mag. (4), 1869, 37, 384; 38, 156.

[41] G. Mie. Ann. Physik (4), 1908, 25, 377.

[42] N.A. Clark, J.H. Lunacek and G.B. Benedek, Amer. J. Physics, 1970, 38, 575.

[43] J. Errera, J.Th.G. Overbeek and H. Sack, J. Chimie Physique, 1935, 32, 681.

[44] H. Freundlich, "Thixotropy", Hermann, Paris, 1935. T. Péterfi, Arch. Entwicklungsmech. Organ., 1927, 112, 689.

2

Attractive Surface Forces

By D. Tabor

PHYSICS AND CHEMISTRY OF SOLIDS, CAVENDISH LABORATORY, MADINGLEY ROAD, CAMBRIDGE, U.K.

Abstract

This paper deals primarily with dispersion forces which are exerted by all atoms or molecules, whether polar or not. They arise from fluctuations in the electron charge distribution around the atom or molecule. These fluctuations produce an instantaneous dipole so that one molecule in the neighbourhood of another experiences an attraction, the potential for a separation of r falling off as r^{-6}. For separations greater than a few tens of nm the fluctuations are in poor correlation and in this regime of "retarded" forces the potential falls off as r^{-7}. The dispersion forces between macroscopic bodies may be calculated from the individual atomic forces by a simple type of pair-wise addition first suggested by Hamaker. A more satisfactory method due to Lifshitz treats each body as a continuum and expresses the interaction in terms of the dielectric properties of the two bodies. This approach is particularly useful if the bodies are immersed in a medium other than air or vacuum and has therefore been of great value in studies of colloidal stability. Broadly speaking the interactions are now well understood for separations from 1 to 10 nm (non-retarded) and from 10 to 1000 nm (retarded). However, both treatments are inadequate in dealing with separations less than 0.2 nm and are particularly misleading in the case of atomic contact. Here two main factors are involved. First the intrinsic graininess of matter; secondly, as orbitals come into contact extra interactive energies are involved which are about twice as large as the calculated dispersion energies. This is of major importance in calculations of surface energies and interfacial energies, in adsorption studies and in the whole field of adhesion.

Introduction

Physical Chemists in general and Surface Chemists in particular think of surface forces primarily in terms of colloids and colloidal stability. However, surface forces embrace much wider concepts and involve scientific challenges in fields that are still relatively neglected. In this article I wish to refer both to the more conventional and to the less conventional aspects: I shall restrict myself primarily to dispersion forces.

van der Waals Forces: their range of action

There are as we shall see, many types of surface forces but the force which is exerted by all surfaces is that due to London (or dispersion) van der Waals interactions. The explanation for this has now been with us for over 50 years (London 1930)[1]. Whereas polar materials clearly exert electrostatic forces on other dipoles (Keesom, 1912)[2] and polar molecules can attract non-polar molecules by inducing dipoles in them (Debye 1920)[3] the attraction between non-polar atoms and molecules remained a mystery until it was realised that the electron cloud surrounding the nucleus could show local fluctuations of charge density. This would produce a dipole moment the direction of the dipole fluctuating with the frequency of the charge fluctuations. Thus if one could view an argon atom for a very short interval of time (of order 10^{-15} to 10^{-16}s) the electron distribution would appear to be more concentrated on one side of the nucleus than the other. This implies that the argon atom has an instantaneous dipole, say μ. This propagates an electric field E which at a distance of x from the atom has a strength of order $(\mu/x^3)(4\pi\epsilon_0)^{-1}$. If there is another atom at this point the field polarises it and creates a dipole μ^1 of strength $\mu^1 = \alpha E$ where α is the polarisability of the atom. The potential energy of u of a dipole μ^1 in a field E is $u = -\mu^1 E$. This leads to

$$u = -k/x^6 \tag{1}$$

London solved the problem for the simplest case of two atoms in which interactions were assumed to occur through their S electrons. These were treated as quantum oscillators of frequency ν and it was assumed that a small displacement of the electrons from their equilibrium position produced instantaneous dipoles which interacted to change the energy of the system. This was fed into the Schrödinger equation as a perturbation and the new zero point energy of the system determined. For two identical atoms of polarisability α, distance x apart, he obtained for the interaction energy u at O K.

$$u = - \frac{3}{4} h\nu\alpha^2 x^{-6} (4\pi\varepsilon_o)^{-2} \qquad (2)$$

The term $h\nu$ was taken as some typical feature of the energy of the individual atom and was generally identified with the ionisation potential. This implies a single frequency of fluctuating charge and is clearly an oversimplification. The frequencies at which the charge fluctuates are those at which the atom or molecule (or bulk materials in the condensed state where individual energy levels are broadened into energy bands) can absorb electromagnetic radiation. A whole range of energies is thus available and each of these will have its own transition probability. Thus one can envisage a wide range of fluctuations of various oscillator strengths. Taking $h\nu$ as the ionisation potential is equivalent to taking the largest single energy transition that is possible, if one ignores any possible contribution from the ionised continuum states.

The London treatment may also be applied to non-identical atoms. For two different atoms A and B of polarisabilities α_A and α_B and of ionisation potentials $h\nu_A$ and $h\nu_B$ respectively the interaction energy u_{AB} is again given by equation (2) with ν replaced by $2\nu_{AB}/(\nu_A+\nu_B)$ and α^2 by $\alpha_A\alpha_B$. It should, however, be noted that the London equation is derived from a model in which only one electron is involved in the fluctuations. In principle all the orbital electrons should be involved. However, those in inner shells have an extremely small probability of taking part in energy transitions so that their contribution to the overall dispersion forces will be correspondingly small. There are many semiempirical relations for estimating the effective number of outer electrons that should be counted in assessing the strength of the resultant dispersion force. To a non-specialist this area still appears to be unsatisfactory.

There are two broad limitations to the London solution even for the simplest atoms or molecules.

<u>Large separations: retarded forces</u>

We have already seen that London's analysis is based on instantaneous dipoles which fluctuate over periods of the order of 10^{-15} to 10^{-16} s. If two atoms are further apart than a certain distance, by the time the electric field from one dipole has reached and polarised the other, the electron configuration of the

first atom will have changed. There will be poor correlation
between the two dipoles and the two atoms now experience what is
known as "retarded" van der Waals forces. This occurs when the
separation x_c is of the order:- fluctuation time τ x velocity c of
the electric field in space. Putting τ = say 10^{-16}s and c =
velocity of light = 3 x 10^8m s^{-1} we have $x_c \approx$ 3 x 10^{-8}m = 300 Å.
The rigorous theory of retarded forces is extremely complicated but
the following simple account (Tabor, 1979)[4] shows how it arises and
why it leads to a different law of force. We start with the London
relation and replace the frequency ν by c/λ where λ is the wavelength of the fluctuation. Then

$$u = -(^3/_4)\alpha^2 h\nu/x^6 (4\pi\epsilon_0)^2 = -\frac{3\alpha^2}{4} \frac{hc}{\lambda x^6} \frac{1}{(4\pi\epsilon_0)} \qquad (3)$$

We now suppose that all wavelengths λ appreciably less than x will
involve fluctuations too rapid at that separation to provide good
correlation. Only those fluctuations which are slow enough to
enable the field to reach a neighbouring atom without much change
in electron configuration will give good correlation and contribute
to the van der Waals interaction. Thus equation (3) will apply
only if λ is equal to or larger than the separation x. In that
case

$$u \approx -\, ^3/_4 \, \alpha^2 \frac{hc}{x^7} (\frac{1}{4\pi\epsilon_0})^2 \qquad (4)$$

This is the "retarded" potential and it is interesting to note
that Casimir gives (in his rigorous solution involving complex
interactions between radiation and electron transitions) the
identical result except for a different numerical constant (Casimir
and Polder 1948)[5]. This simple derivation emphasises two points.
First, retarded forces are not a "new" type of force. They arise
in the same way as the non-retarded forces except that they reject
high-frequency fluctuations when the atoms are more than a few
hundred angstroms apart. (If the velocity of electromagnetic
fields were infinite there would be no retardation!) Secondly
for separations greater than this distance the system will always
ignore the faster fluctuations and will pick out those fluctuations
slow enough to give good correlation.

There is a limit to this type of argument which is shown if
we consider the effect of temperature. The London and Casimir
treatments refer to the zero point energy of the system i.e. the
behaviour at O K. At finite temperatures the system will also
have vibrational, rotational and other forms of thermal energy. Of

these the most rapid is the vibrational frequency which in molecules and in solids is of the order $\nu = 10^{12}$ to 10^{13} Hz. If the
vibrating system carries an intrinsic dipole these fluctuations
will generate an electric field with this frequency say $\nu = 10^{13}$ Hz.
Such a field can travel a distance c/ν i.e. 3×10^{-5}m or 300 000 $\overset{o}{A}$
in the time of a single vibration. Consequently for any separation
smaller than this the van der Waals interaction due to thermal
vibrations will be non-retarded. Thus at these large separations
where the contributions from the more rapid electronic fluctuations
have died away as retarded forces (because the potential falls off
as r^{-7}) the remaining forces will be non-retarded. If rotations
of (polar) molecules are possible the frequency is even lower so
that non-retarded forces will tend to persist at even larger
separations. Of course the magnitude of the forces at this stage
will be extremely small. We may summarise this discussion as
follows:- in general we expect relatively strong non-retarded
forces for separations less than say 300 $\overset{o}{A}$, retarded forces for
separations between say 1000 and 10000 $\overset{o}{A}$ and extremely weak non-
retarded forces at separations greater than this.

<u>Very short separations - atomic contact</u>

The London derivation implies that the field produced by a
dipole μ at a distance x is of the form μ/x^3. This is true only if
x is fairly long compared with the length of the dipole. The
latter is smaller than, but of the same order of magnitude as,
the atomic diameter. It follows that when the separation becomes
comparable with the diameter of the atoms or molecules the simple
equation for the field is no longer valid. The error will be most
marked when atomic contact occurs.

There are two ways of dealing with this. The classical method
favoured by chemists is to recognise that the interacting atoms
can no longer be regarded as simple dipoles: higher orders of
polarisation must be considered and the atoms must be treated as
quadrupoles and octupoles. The potential of these multipoles falls
off more rapidly than dipoles so that their contribution is important only at short separations. The main difficulty here is the
problem of allocating the correct strengths to these higher orders
of polarisation: it involves a combination of theory and empiric-
ism. Nevertheless, in many cases a close approximation to the
correct interaction energy may be achieved (Buckingham 1978)[6]. The
second approach is that due to Gordon and Kim (1972)[7] (for a review
seen Clugston 1978)[8] and tends to be more favoured by physicists.

The electron density is assumed to be the sum of the original
densities of the unperturbed atoms, and the Coulombic, kinetic,
exchange and correlation energies are calculated as a function of
atomic separation by standard methods. The most unsatisfactory
feature of this approach is the crude oversimplified nature of the
basic assumptions. In spite of this shortcoming the calculated
results appear to be in reasonable agreement with the observed
interaction energies as deduced, for example, from scattering
experiments and the results are close to those obtained by the
multipole model. A stylised representation of the behaviour of two
argon atoms as they are brought together is shown in Figure 1. It

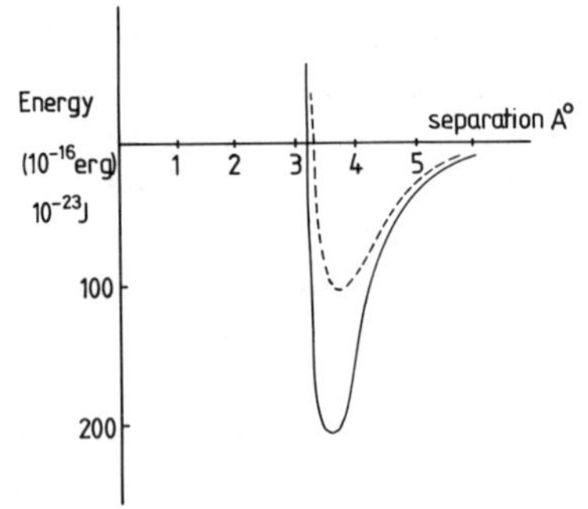

is seen that when atomic contact
occurs the (negative) potential
energy is almost twice as big as
that deduced from the simple
London relation.

This discrepancy must be of
major importance whenever atomic
contact occurs, for example in
the adsorption of a vapour on a
solid surface, at the interface
between two liquids or a liquid
and a solid or between two solids
(as in coagulation of colloidal
particles). Detailed calculations of adhesion or interfacial
energies based solely on a London

Figure 1. Interaction energy
between two argon atoms.
-----simple van der Waals dispersion energy (r^{-6}) with repulsive potential (r^{-12}).
_____including short-range forces
(Gordon and Kim)[7]

type of equation may give answers of the right order of magnitude
but numerically will inevitably be in considerable error.

Forces between macroscopic bodies: the dispersion component

The forces between two bodies due to dispersion effects are
generally referred to as surface forces. Of course they are due
not only to atoms on the surface (see however below) but to atoms
in the hinterland since each atom in one body will exert dispersion
forces on each atom in the other. This leads to a very simple
type of computation (Hamaker 1937)[9] in which the forces between the
bodies are calculated by pair-wise addition. Instead of using the
potential $u = -c/x^6$ for each atomic pair the computation becomes
much easier if the individual atoms are replaced by a smeared out
uniform density of matter such that a small volume Δv_1 exerts, on a
volume Δv_2 in the neighbouring body, a potential given by

$-(c/x^6)n_1\Delta v_1 n_2 \Delta v_2$ where n_1, n_2 represent average numbers of atoms per unit volume in each body. The addition then becomes an integration. We quote two well known examples, a sphere of radius R distance H from a flat surface, and two semi-infinite plane surfaces at distance H apart. The bodies are of the same material containing n atoms per unit volume and the intervening space is vacuum (or air).

Table 1. Forces due to dispersion effects

Force	Potential	Force Sphere on flat	Force/area Flat on flat	Hamaker constant
Unretarded	$-c_1/x^6$	$\dfrac{AR}{6H^2}$	$\dfrac{A}{6\pi H^3}$	$A=\pi^2 c_1 n^{2}$*
Retarded	$-c_2/x^7$	$\dfrac{2\pi BR}{3H^3}$	$\dfrac{B}{H^4}$	$B=0.1\pi c_2 n^2$

* If two different materials are involved $A_{12} = \pi^2 c_{12} n_1 n_2$

<u>Flat parallel surfaces</u>. There are four consequences that may be drawn from these results. i) For flat parallel half spaces the force per unit area is $A/6\pi H^3$ so that the potential energy per unit area $U_\infty = -A/12\pi H^2$. We may now consider the way in which the interaction energy falls off as the half spaces are reduced to slabs of finite thickness. Suppose we consider two parallel slabs of thickness H separated by a distance H. Following the procedure used by Hamaker we obtain (see Figure 2).

$$u = \frac{11}{18}(A/12\pi H^2) = 0.61\, u_\infty$$

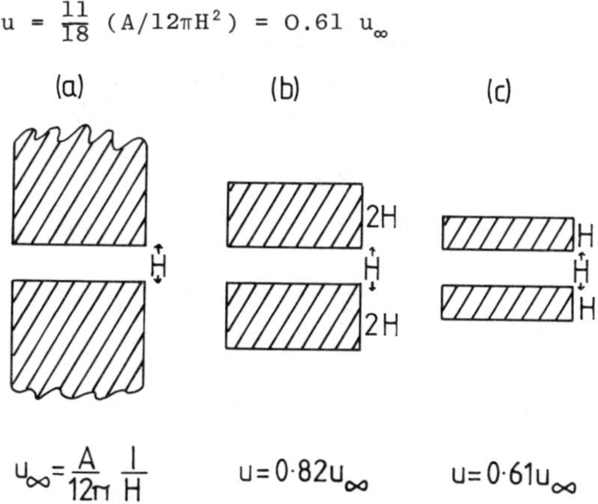

Figure 2. Interaction energy u per unit area for parallel surfaces distance H apart.
(a) Semi-infinite solids.
(b) Solids of thickness 2H.
(c) Solids of thickness H.

In the limit if H is made equal to the atomic spacing z we have

$$u/\text{unit area} = 0.61\ A/12\pi z^2 \qquad (5)$$

The force is the differential of this. We are thus left with the conclusion that two parallel planes, only one atom thick and one atom apart exert a force on one another that is over 60% of the force between two semi-infinite slabs. For atomic contact it is not obvious what value we ought to assume for the effective separation: it clearly lies between O and z. If we take a rough value of z/2 as the effective separation the energy increases from 61% of u_∞ to 82% of u_∞. Thus the superficial atoms contribute appreciably more than half the interaction of two semi-infinite slabs and this effect will be accentuated if there is screening of the field by the electrons in the outer atomic layer* ii) In view of the fact discussed in a preceding section that the van der Waals dispersion forces are substantially increased when the orbitals are in contact this further emphasises the importance of the interfacial atoms as distinct from the contribution of those in the hinterland. iii) If atomic contact occurs one can no longer use the Hamaker model which treats the solid as a smeared-out uniform material. The "graininess" of matter reappears and is particularly important for the surface atoms. Consider for example a simple cubic array of atoms for which the atomic separation is z. We allow only for the interaction between nearest neighbours. Then for two neighbouring atomic planes the potential energy per unit area is $-(c/z^2)$ x $(1/z^2)$ since there are $(1/z^2)$ atoms per unit area. Hence

$$u/\text{unit area} = -\ cz^{-8} \qquad (6)$$

But the Hamaker constant $A = \pi^2 c\ n^2 = \pi^2 cz^{-6}$. Equation (6) then becomes

$$u/\text{unit area} = -A/(\pi^2 z^2) \simeq -A/(10z^2) \qquad (7)$$

If this is compared with the result for two semi-infinite surfaces putting H = z

$$u/\text{unit area} = -A/(12\pi H^2) \simeq -A/(40z^2) \qquad (8)$$

---

* This is consistent with the classic observation (of Corkhill, Goodman, Ogden & Tate 1962)[10] that the surface tension (free surface energy) of a film of water only 20 Å thick is indistinguishable from that of bulk water. Of course in this case hydrogen bonding plays a dominant part. Again Requena, Billett & Haydon (1975)[11] found that films only 50 Å in thickness, formed under aqueous solutions from decane containing a small amount of lipid, have an interfacial tension very close to that of bulk hydrocarbon solution.

The two monolayers actually give a larger reaction energy than the semi-infinite surfaces. Better agreement would result if we assumed that $H \simeq z/2$ at contact. Evidently the Hamaker equation is unreliable when one deals with atomic contact though it will give answers of the right order of magnitude. iv) An approximate relation for the interaction energy between materials A and B may be readily deduced. If each material has atomic polarisability α_A, α_B respectively and if the resonant frequencies are approximately the same, we have from London's equation

$$u_{AA} = - c_{AA}/x^6 \quad \text{where} \quad c_{AA} = 3/4 \ h\nu\alpha_A^2(4\pi\epsilon_0)^{-2} \qquad (9a)$$

$$u_{BB} = - c_{BB}/x^6 \quad \text{where} \quad c_{BB} = 3/4 \ h\nu\alpha_B^2(4\pi\epsilon_0)^{-2} \qquad (9b)$$

$$u_{AB} = - c_{AB}/x^6 \quad \text{where} \quad c_{AB} = 3/4 \ h\nu\alpha_A\alpha_B \ (4\pi\epsilon_0)^{-2} \qquad (9c)$$

$$\text{Thus} \ c_{AB}^2 = c_{AA} \ c_{BB}$$

$$(10)$$

$$\text{Similarly} \ A_{AA} = \pi^2 n_A^2 c_{AA} \ : \ A_{BB} = \pi^2 n_B^2 c_{BB}$$

$$\text{Thus} \quad A_{AB} = \pi^2 n_A n_B c_{AB} = (A_{AA} \ A_{BB})^{\frac{1}{2}} \qquad (11)$$

This is the basis of all the approximate relations for the interaction between two different materials. We may take this one stage further and express the results in terms of surface energies. Ignoring the complications mentioned above when atomic contact occurs we may calculate the surface energy of a van der Waals solid by noting that the work required to separate two unit areas from atomic contact to infinity is $A/(12\pi z^2)$. This is the energy required to produce two new unit surfaces* and is thus equal to 2γ Hence

$$\gamma = A/(24\pi z^2) \qquad (12)$$

As a matter of interest we quote results for hydrocarbon multi-layers deposited on mica substrates (Coakley and Tabor 1977)[12] for which it is found directly that $A = 10^{-19}$ J. Then if $z \simeq 2 \ \overset{\circ}{A} = 2 \times 10^{-10}$ m we obtain $\gamma \simeq 30$ mJ m^{-2}. This may be compared with the general observation that hydrocarbons are generally wetted by

---

* In this and all similar discussions in this paper we have ignored the repulsive potentials.

liquids of surface energy less than 30 mJ m^{-2} so that the solid
surface energy must also be of this order of magnitude. Clearly
the calculated value of γ depends rather critically on the assumed
value of z. For this reason we must view rather critically the
value for γ we have quoted above as well as the excellent agreement obtained by other workers (e.g. Fowkes,[13]; Padday and
Uffindell[14]) using the simple Hamaker expression to determine
interfacial energies and surface energies of various liquids.

Using equation (12) we have

$$\gamma_{AA} = A_{AA}/(24\pi z_{AA}^2) \; : \; \gamma_{BB} = A_{BB}/(24\pi z_{BB}^2) \; : \; \gamma_{AB} = A_{AB}/(24\pi z_{AB}^2)$$

(13)

If the atomic separations z_{AA}, z_{BB}, z_{AB} are all comparable (more
specifically if $z_{AB}^2 = z_{AA} \cdot z_{BB}$) we end up with the relation for the
surface energies.

$$\gamma_{AB} = (\gamma_{AA}\gamma_{BB})^{\frac{1}{2}}$$

(14)

This type of relation has been widely used though it is clear from
all the preceding approximations and assumptions that it must again
be regarded as an order of magnitude calculation, not a rigorous
derivation. In general the Lifshitz approach is much more reliable
(see below). However, because it treats each body as a continuum
with specified bulk properties it also becomes unreliable at atomic
contact where the graininess of matter becomes evident. Furthermore calculations for the interfacial energy of systems consisting
of molecular (as distinct from monatomic) liquids are likely to
involve additional errors since the molecules may acquire some
specific orientation at the interface.

Sphere on flat. The sphere on flat gives a well defined result
and measurements can be made down to very small separations. This
would be experimentally impossible with parallel flat surfaces.

Forces between macroscopic bodies: the continuum approach

The method of pair-wise addition due to Hamaker has three
main defects. First it is evident that if atom 1 in A exerts a
force on atom 1 in B the presence of neighbouring atoms in A and
B is bound to influence the interaction between A1 and B1.
Secondly it is not obvious whether the electric field propagated
by an instantaneous dipole some distance from the free surface
ought to take into account the finite dielectric properties of the
material that lies between it and atoms in the other body. Thirdly
it is not clear how the addition needs to be modified if the space

between bodies A and B is not air or vacuum but another dielectric
medium. Hamaker attempted to deal with the latter point by sub-
tracting a sort of buoyancy effect due to the intervening material.

The most radical solution to all these problems is that due
to Lifshitz who removed all the difficulties by treating the bodies
and the intervening medium in terms of their <u>bulk</u> properties,
specifically their dielectric constants. The Lifshitz approach
has been discussed in a number of text books and review articles
at various levels of sophistication (Israelachvili and Tabor
(1973)[15], Israelachvili (1974)[16], Langbein (1974)[17], Mahanty and
Ninham (1976)[18], Margenau and Kestner (1971)[19], Parsegian (1975)[20]).
If we consider two semi-infinite parallel bodies of dielectric
constants ε_1 and ε_2 separated by a medium of dielectric constant
ε_3 the interaction energy is ultimately expressed in terms of
$\frac{\varepsilon_1-\varepsilon_3}{\varepsilon_1+\varepsilon_3}$ and $\frac{\varepsilon_2-\varepsilon_3}{\varepsilon_2+\varepsilon_3}$. These quantities turn out to be equivalent to the
effective polarisabilities of these bodies in the presence of the
intervening material.

The Lifshitz approach shows, amongst other results that if
$\varepsilon_1<\varepsilon_3<\varepsilon_2$ bodies 1 and 2 repel one another. This is because they
attract the intervening medium more than they attract one another.
However, if $\varepsilon_1 = \varepsilon_2$ there is always attraction whatever the
value of ε_3. Thus a film of liquid with air on both sides
($\varepsilon_1 = \varepsilon_2 = 1$) will undergo continuous thinning due to its "own"
van der Waals forces.

The Lifshitz approach is particularly suited for macroscopic
bodies for separations greater than a few Ångstroms. For smaller
separations there are two complications - first the dielectric
properties are not the same at the surface as in the bulk -
secondly the graininess of matter may introduce important perturb-
ations. Thus the Lifshitz approach is not reliable when atomic
contact is involved. However, it has been used very skilfully by
Israelachvili (1973)[21] to calculate the surface energies and contact angles of various non-polar organic liquids: but the remarkable
agreement between theory and experiment must be viewed as being
partly fortuitous. Apart from these limitations when atomic
contact occurs the Lifshitz treatment deals very successfully with
bodies made of layers of different dielectric properties. It is
also successful in coping with the problem of solids separated
by a liquid dielectric. Further there are now detailed analyses
for the behaviour of two solids in the presence of a liquid

electrolyte[22].

For simple geometries the Lifshitz results for bodies in air do not differ markedly from those provided by the Hamaker analysis. For example the force between two spheres each of radius R is of the form $F = AR/(6 H^2)$ where A is not greatly different from the conventional Hamaker constant. This is for separations between say 10 and 300 $\overset{\circ}{A}$. For larger separations the Lifshitz analysis leads naturally to a regime of retarded interactions, giving a force proportional to R/H^3. However, for separations greater than say 1000 $\overset{\circ}{A}$ the Lifshitz treatment shows that thermal fluctuations become increasingly important and the force law begins to tend, once more, to R/H^2. For very small particles Langbein suggests that the forces fall off as $1/H^7$ as though the particles were single atoms.

The direct measurement of surface forces as a function of separation

So long as we are not concerned with intimate atomic contact, surface forces are generally the result of van der Waals dispersion forces. Macroscopic measurements of this type have been mainly carried out, in air, between a spherical surface of radius R and a flat surface or between crossed cylinders each of radius R - this is equivalent to a sphere on a flat. One expects to find either non-retarded behaviour of the type $F = AR/6H^2$ or retarded behaviour of type $F = 2\pi BR/3H^3$. The early classical work by the Dutch and Russian schools involved the use of polished quartz surfaces. Most of their results were restricted to separations greater than several hundred angstroms and they confirmed the existence of forces which fell off as H^{-3} i.e. retarded forces. With the use of mica surfaces (which are molecularly smooth) arranged as crossed cylinders it was possible to measure forces down to separations of the order of 20 $\overset{\circ}{A}$ or less. Typical results taken from the papers of Tabor and Winterton[23] and Israelachvili and Tabor[24] are shown in Figure 3, from which it is possible to deduce that non-retarded forces operate for separations less than say 50 $\overset{\circ}{A}$ and retarded forces for separations greater than about 500 $\overset{\circ}{A}$ (see Figure 4). Another very interesting method is due to Anderson and Sabinsky (1970)[25] who measured the shape and thickness of the meniscus of a film of liquid helium in contact with a glass wall. This is especially relevant to the short-range unretarded forces at the interface.

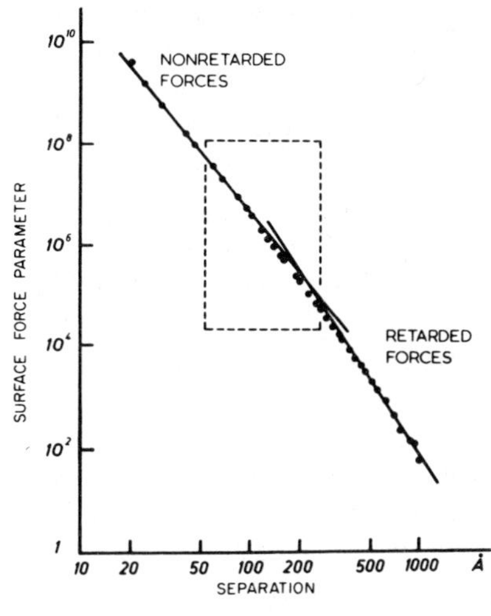

Figure 3. Experimental results (taken from Israelachvili and Tabor[15,24]) showing the variation of force between two curved mica surfaces as function of separation. The force parameter is such that the slope has a gradient (n + 2) where n is the power law dependence of the force. The region enclosed by broken lines is the experimental region covered by the earlier work of Tabor and Winterton[23].

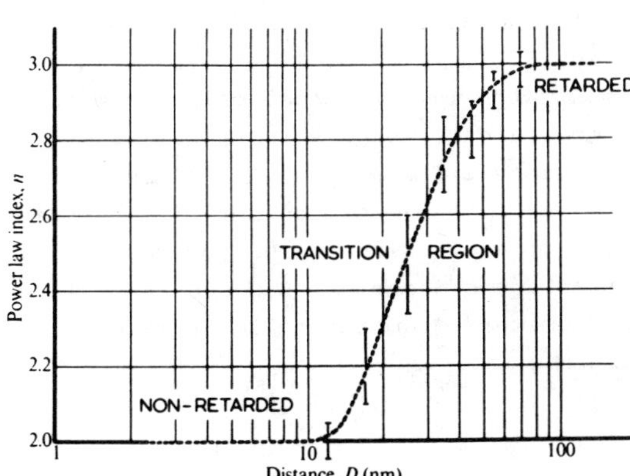

Figure 4. Variation of the power law n of the van der Waals law of force between crossed mica cylinders with distance D. The curve is based on the combined results of a number of jump and resonance experiments. (Israelachvili & Tabor)[24]

<u>The calculation of van der Waals forces</u>

Before leaving the experimental results obtained with the mica surfaces it is of interest to consider the attempts to calculate the dispersion forces to be expected in terms of the known dielectric constant at all the relevant frequencies. Different interpolation and extrapolation procedures can lead to substantially different results. For example Figure 5 shows the two sets of theoretical results obtained by Richmond (Tabor (1977)[26], Chan and Richmond (1977)[27]) from the same dielectric data using different interpolation procedures. The differences are very marked, the second set agreeing reasonably well with experiment. Of particular interest is the tendency for the law of force to change over to

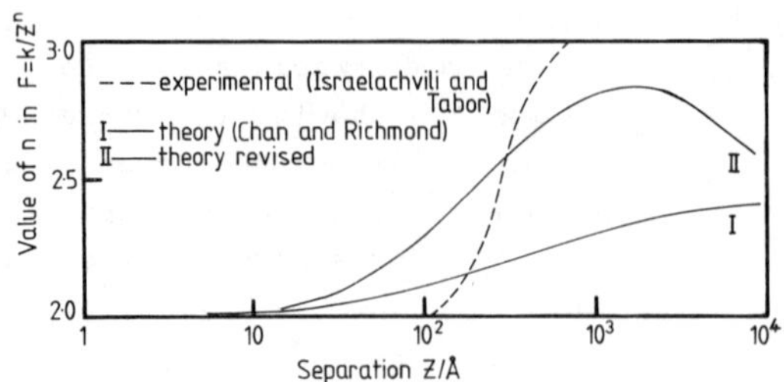

Figure 5. Force separation curve for mica surfaces in air showing transition from non-retarded (n=2) to retarded forces (n=3).
---- Experimental: Israelachvili & Tabor[24].

I _____ Theory: Chan & Richmond; unpublished (see Tabor)[26].
II ——— Revised theory: Chan and Richmond[27].

non-retarded at very large separations. This is because in this regime the behaviour is dominated by thermal fluctuations (at temperatures above O K). This follows from the Lifshitz approach.

Surface forces in the presence of a liquid dielectric

Israelachvili has recently described some original, beautiful and extremely interesting results for mica surfaces separated by a liquid dielectric. The most striking are those obtained with a non-polar liquid where the molecule is roughly spherical and about 10 Å diameter (Horn and Israelachvili, 1980)[28]. If all moisture is extracted from the system the force as a function of separation is as shown in Figure 6. i.e. it shows fluctuations corresponding to a layered structure in the liquid which extends over a distance of the order of 6-8 atomic diameters from the solid surfaces. The

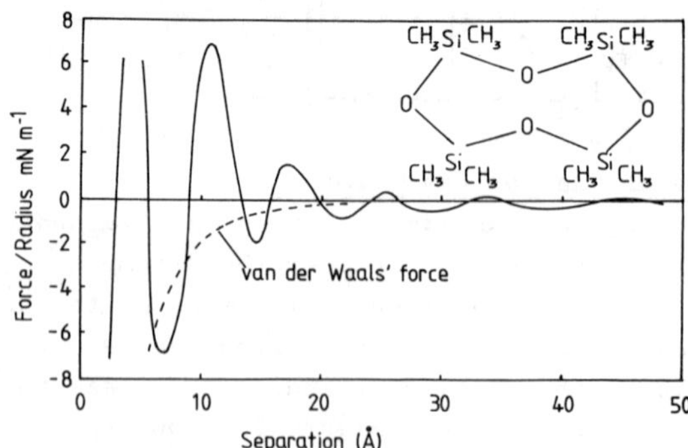

Figure 6. Force F between curved mica surfaces (effective radius of curvature R) immersed in OMCTS (octamethylcyclotetrasiloxane)[28]. The molecules of the liquid are roughly spherical with a diameter of about 10 Å.

liquid is "trapped" between the two mica surfaces and at non-integral molecular separations the liquid molecules are less closely packed than if the liquid were unconstrained. This gives a density change and a corresponding change in the force between the solid surfaces. If the liquid is allowed to retain a small amount of moisture the fluctuations disappear. This may be because the water is adsorbed preferentially on the mica surface thus reducing the strength of attachment of the first layer and this may destroy the structuring. However, it should be noted that the mica surface, being prepared in air, probably already carries a monolayer of water. The details are still the subject of discussion.

If the liquid between the surfaces is water no structural features are observed (Israelachvili, private communication): this may be because the water molecule is too small but the matter is still under investigation. On the other hand structural features are rather marked with nematic liquids where the structural length is quite large.[29] These highly original observations are of very great interest not only in the field of surface forces but also in relation to the structure of liquids.

<u>Surface forces and adhesion</u>

Colloid scientists have long been interested in the adhesion between colloid particles should they come into molecular contact. For this reason most of the systems studied have been between crossed cylinders or between spheres or between a sphere and a flat. We consider here a sphere on a flat of the same material. The surfaces are assumed to be molecularly smooth so that the geometric area of contact can be considered to be equal to the atomic area of contact.

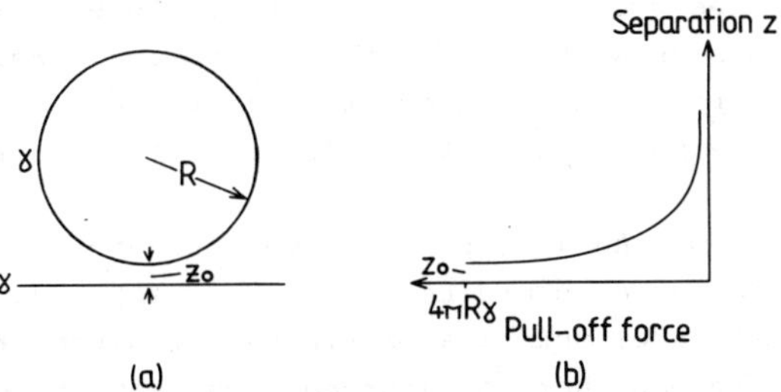

Figure 7. Surface Forces & the adhesion between a rigid sphere of radius R and a rigid flat.(a) Situation at zero applied load. (b) Pull-off force Z as function of separation z. The maximum of Z occurs when the bodies are in atomic contact (separation z_0) and has the value $Z = 4\pi R\gamma$.

                                                          Colloidal Dispersions

Ideal rigid solids Figure 7a. No deformation of the bodies occurs.
If the surfaces are atomically smooth the distance of nearest
approach is the atomic spacing z and the dispersion attractive
force is

$$F = AR/6z^2 \qquad (15)$$

Using equation (12) to eliminate A/z^2 we obtain

$$F = 4\pi R\gamma \qquad (16)$$

Thus the force to pull the surfaces apart will be a maximum when
the surfaces are in contact and will diminish as the separation
is increased. (Figure 7b).

Derjaguin modified the above treatment to allow for elastic deformation of the bodies (Figure 8a). In his model the sphere,

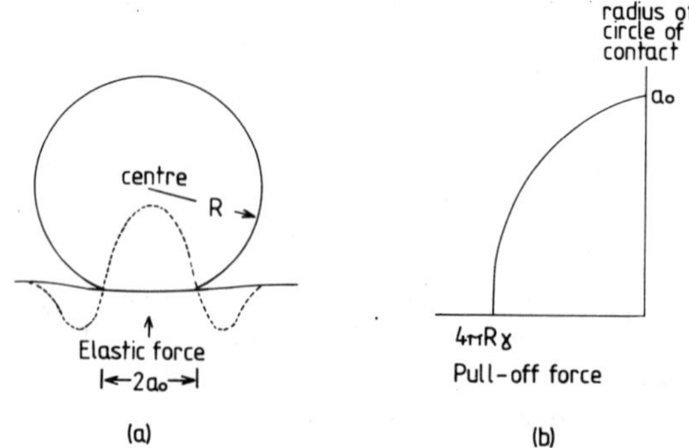

Figure 8. Surface forces and adhesion between a sphere of radius
R and a flat assuming classical Hertzian elastic deformation in the
contact zone (Derjaguin model)[30]. (a) Contact zone under zero
applied load: the broken line shows the pressure distribution
which is compressive within the contact zone and tensile outside.
(b) Pull-off force which reaches its maximum value $4\pi R\gamma$ when the
contact radius is zero, i.e. at atomic contact as in Figure 7a.

under the action of the attractive forces, penetrates the flat
and is itself deformed according to the Hertzian equation for the
elastic deformation of spherical surfaces: a circular contact
of radius a is formed where a is proportional to the one third
power of the normal force (Derjaguin et al. 1975)[30]. This process
has two consequences. First because the circle of contact is now
in internal equilibrium with repulsive forces of atoms it contri-
butes nothing to the attractive force. This is now provided only

by the surfaces outside the circle of contact. Secondly, the
elastic forces resisting penetration increase as a increases –
a limit is reached at a_o when the forces balance and no further
penetration occurs. If now a pull-off force Z is applied the
circle of contact gradually diminishes until the whole of the
elastic deformation has recovered and the sphere and flat surface
are in undeformed contact. At this stage Z has reached its maximum value $4\pi R\gamma$ and at this force the surfaces will pull apart
(Figure 8b). This analysis was criticised on the grounds that the
deformation of the surfaces will not be that given simply by the
Hertzian equations – the Hertzian deformation must be modified to
take into account the surface forces themselves. This is most
clearly revealed if one studies the behaviour of highly elastic
solids.

<u>Rubber sphere against a rigid flat surface</u>. For simplicity we
assume that both surfaces have the same surface energy γ and the
interfacial energy is zero. (If the surface energy of the sphere
is γ_1, of the flat γ_2 and of the interface γ_{12} this is equivalent
to replacing the "adhesion" energy 2γ by $\gamma_1 + \gamma_2 - \gamma_{12}$). An analysis
of this problem was first given by Johnson, Kendall and Roberts
(J.K.R.) (1971)[32]. It is described most simply in the following
terms. If the surfaces are placed in contact under zero load, the
area of contact according to the Hertzian equation should be zero.
However, surface forces pull the surfaces together and form a neck
as shown in Figure 9a providing a circle of atomic contact of

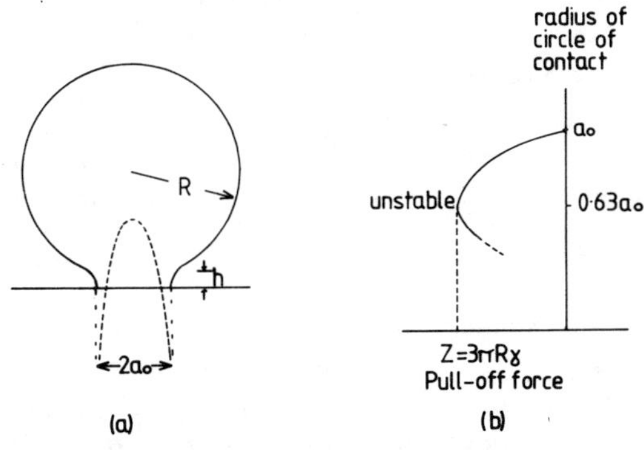

Figure 9. Surface forces & adhesion between an elastic sphere of radius R & a hard flat assuming that the surface forces are negligibly small outside the zone of contact (JKR model)[32].

(a) Contact zone under zero applied load: the broken line shows the pressure distribution which is compressive at the centre and tensile at the edges of the contact zone.

(b) Pull-off force as function of radius a of circle of contact.
Instability occurs when $a = 0.63a_o$ at a pull-off force equal to
$3\pi R\gamma$.

radius a_o. This releases surface energy of amount $2\gamma A$ where A is
the area of contact and equals πa_o^2. This energy is stored as
elastic strain energy in rubber. If a pull-off force is applied
it will reduce the area of the circle of contact until a point is
reached where a very small decrease in A, say ΔA, will release
more elastic strain energy than is necessary to provide the requi-
site surface energy $2\gamma.\Delta A$. This is a point of instability and, at
the critical pull-off force at which this occurs the surfaces will
pull apart (Figure 9b). The behaviour is exactly like that of
crack propagation and indeed one way of studying the problem is
to treat it by the methods of fracture mechanics. Pull-off occurs
for a force $Z = 3\pi R\gamma$ and the circle of contact at which this occurs
has a radius a given by

$$a \simeq (\tfrac{1}{4})^{\frac{1}{3}}\, a_o = 0.63\, a_o \qquad\qquad\qquad (17)$$

Experiments with rubber show that this instability does indeed
occur at pull-off and even with a relatively stiff material such as
mica a similar instability is observed (Tabor,[26]; Israelachvili,[31]).
      The main defect of this model is the following. The whole
of the surface energy change is assumed to be restricted to the
contact region. This is a fair assumption for macroscopic bodies
but if the height of the neck h (see Figure 9a) is comparable with
atomic dimensions there will be a contribution to the surface
interactions from material outside the immediate contact region.
Consequently larger attractive forces are to be expected. In a
discussion of this feature (Tabor 1977)[26] I pointed out that $h \approx$
$(R\gamma^2/E^2)^{\frac{1}{3}}$ and that therefore we might expect a deviation from the
J.K.R. solution when $h \approx 1$ to $3\ \overset{o}{A} \approx z_o$ where z_o is the atomic
diameter. Thus the dimensionless parameter marking the transition
between the Derjaguin and the J.K.R. model would be of the form

$$\emptyset\, (^h/z) = \phi\, (R\gamma^2/E^2 z_o^3)^{\frac{1}{3}} \qquad\qquad\qquad (18)$$

      The (1977) paper led to an exchange of correspondence in the
Journal of Colloid and Interface Science between Professor Derjaguin
and myself. Three years later two of Prof. Derjaguin's colleagues
published a paper which, in fact, provides a very satisfactory
solution to the problem. Probably better solutions will follow but
the main conclusions will not be changed. The authors (Muller,
Yushchenko and Derjaguin, 1980)[33] show that for very rigid surfaces
the pull-off force will be equal to $4\pi R\gamma$ while for highly elastic

solids it will approach $3\pi R\gamma$. The crucial parameter determining this
transition is almost identical with that given in equation (18).
It is clear that there is now general agreement about the adhesion
process for spherical surfaces attracted by van der Waals forces.
There still, however, remains some uncertainty as to whether
colloidal particles in practical colloidal systems will show the
"necking" characteristic or not. It depends on the size of the
particles, the modulus and the strength of the interfacial forces
i.e. the Hamaker constant.

Of course the adhesion may be determined by additional factors.
One factor which has been greatly emphasised by Derjaguin is the
electrostatic attraction which occurs at the interface of dissimilar materials as a result of charge transfer (Deryagin, Krotova
and Smilga, 1978)[34]. There is no doubt that there are numerous
situations where they are small or negligible - for example in the
contact between two bodies of the same material. Thus two clean
strips of the same rubber will stick together primarily by van der
Waals forces*. However, the energy expended in pulling the strips
apart involves another factor of considerable importance, namely
hysteretic or visco-elastic losses in the solids. This has
generally been ignored by surface chemists.

Adhesion of visco elastic solids

With materials such as rubber or polymers it is convenient to
consider adhesion in terms of fracture mechanics. For example, for
ideal solids the energy involved in reducing the area of contact
between two surfaces by amount dA i.e. opening up the crack by
amount dA is simply 2γ dA. In the case of ideally elastic
spherical surfaces this is the quantity used in the analysis of
adhesion mentioned above. With real materials, however, energy is
dissipated at the "tip of the crack" where the surfaces separate.
This is because the stress field at this region involves relatively
large strains, and for a typical rubber or polymer this can lead
to large energy losses. (With metals the stress field generally
produces plastic flow at the crack tips and this accounts for the
large amount of energy involved in the rupture of metal joints.)
A detailed analysis by Greenwood and Johnson (1981)[36] shows that
for typical viscoelastic solids this additional energy loss is
proportional to 2γ dA but is very much larger, often 100 or even

---

* We assume that the contact is not maintained for a protracted
period so that interdiffusion across the interface (Voyutski,
1963)[35] can be neglected.

10000 times bigger. Further, experiments by Maugis and Barquins (1978)[37] and Johnson's recent analysis have shown that the pull-off force varies with temperature and rate of pull-off in a way that agrees quantitatively with the visco-elastic properties of the solid. With such a system where chemical bonds are neither formed nor broken the effect of rate of pull-off cannot be attributed to the rate of leaking away of interfacial charge (pace Derjaguin)[34].

The energy balance for viscoelastic solids is coped with in fracture mechanics by writing:-

Fracture energy = energy release per unit area of crack formed
                = "effective" surface energy
                = 2γ + viscoelastic work

where γ is the true equilibrium surface energy. We are familiar with such effects in the behaviour of 'Scotch' tape. The surface energy γ of the adhesive material is of the order of $30m\ Jm^{-2}$, but on account of the enormous amount of viscous energy dissipated as the tape is peeled away the effective value is near $100J\ m^{-2}$. The adhesion literature and particularly the fracture literature contains numerous examples of this type of process. These losses swamp the true thermodynamic surface energy.

<u>Surface forces of other solids</u>

<u>Metals</u>.[38,39] Metal surfaces more than 10 Å apart experience attractive forces which resemble classical dispersive forces in that they arise from fluctuating electron movements within the plasma. The effective Hamaker constant is perhaps five times greater than that for van der Waals solids so that, where atomic contact between two metals occurs the interactive energy due to this source is of the order of $100mJ\ m^{-2}$. However, when the surfaces are about 2 Å apart the metallic bond suddenly forms and this contributes about ten times as much interaction energy (Figure 10). Consequently the surface energy of metals is of the order of 1 to 2 $J\ m^{-2}$. Thus with gold colloidal particles the interaction energy can be relatively high and lead to strong adhesion. There is some evidence that the forces may be large enought to produce plastic deformation of the particles but for this to occur they must be very clean. In practice most metals are covered with oxide films which may be between 50 and 100 Å thick and these completely destroy the metallic bond. On the other hand the oxides themselves will exert their own van der Waals forces and

when atomic contact occurs the closed orbital interactions[7,8] may
be relatively strong.

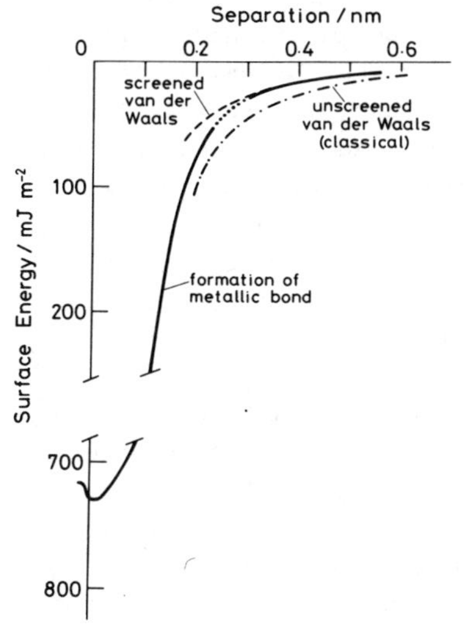

Figure 10.   Interaction energy
between two semi-infinite slabs of
"jellium" as they are brought into
atomic contact (Inglesfield)[39].

Covalent Solids. With
particles made of such materials the main interaction
arises from dispersion forces.
When contact occurs it is a
moot point as to whether
covalent bonds can form
across the interface. If
they can the bonding will be
very strong and the particles
will not easily separate.
Here again it is probable
that the surface of a covalent solid is itself covered
with some chemically formed
contaminant film.

Ionic Solids. If the
electronic charges are uni-
formly distributed over the
free surface the fields due to
positive and negative charges
will tend to cancel one
another and at a separation
of the order of 20 Å the net

field can be very small indeed. In that case dispersion forces
will dominate. When the surfaces come into atomic contact strong
electrostatic forces will come into operation and these may be
augmented if there are residual dipoles in the bulk of the
material.

Conclusions

     All solids exert forces on other bodies due to the presence
of fluctuating charge distributions within them. These dispersion
forces are well understood and in some cases have been measured
directly. For particles in air (aerosols) the forces may be
calculated by pairwise addition (the Hamaker approach) and the
calculated Hamaker constant is in reasonable agreement with
experiment. The forces are non-retarded for separations less than
say 10 nm, retarded for separations greater than 50 nm.

     In the presence of an intervening medium these attractive
forces are usually considerably reduced. The Hamaker approach is

unsatisfactory and the forces must be calculated in terms of the bulk dielectric properties of the particles and of the intervening material using the Lifshitz continuum approach. This is not always easy since the theory requires a knowledge of the dielectric constants over a wide range of frequencies. The calculation in turn depends on the type of procedures used in interpolating or extrapolating from known data. Nevertheless the broad outlines are well established. In the presence of an aqueous medium the interparticle forces are reduced by a factor of about 10. In principle the transition from non-retarded forces should occur at smaller separations than in air due to the reduced velocity of electromagnetic radiation in the medium, but there is little experimental evidence for this.

In the presence of a liquid, fluctuations may be observed in the interparticle force as the particles are brought together at separations less than say 50 $\overset{o}{A}$. This appears to be due to structural features in the liquid even for non-polar materials such as OMCTS (Octamethylcyclotetrasiloxane). With water structural effects are not observed though steric factors must presumably affect interparticle forces for sufficiently small separations.

Most colloidal systems are concerned with particles several tens or hundreds of nm apart. For these systems the interparticle forces can be calculated with some reliability. However, if the particles come into atomic contact, as occurs for example when coagulation takes place, the theories will generally underestimate the strength of the interparticle forces. The Lifshitz approach is also unsatisfactory since it treats each material as a continuum whereas the atomic "graininess" becomes important precisely when contact occurs.

The short-range interactions at atomic contact need more detailed study for van der Waals solids, metals, and other types of solids. Although these forces may not be of major importance in colloidal systems, as we have seen they play a fundamental part in surface energies and in adhesion. Indeed they have now become an important and challenging theme in the Physics and Chemistry of Surfaces.

<u>Acknowledgements</u>

I am grateful to Professor Buckingham and Dr. A.J. Rennie for helpful comments on the manuscript. I also wish to thank Dr. Roger Haydock and other members of the Theory of Condensed Matter (T.C.M.) group, Cavendish, for bringing to my attention the work of Gordon

and Kim. Finally I wish to record that since writing this manuscript I have seen the article by B.W. Ninham entitled "Long-
Range vs. Short Range Forces. The Present State of Play", 1980,
J. Phys. Chem. 84, 1424-1430. This deals primarily with solvent-
mediated forces at distances less than about 30 $\AA$. The emphasis
is thus rather different from that made in the present paper which
deals with solids in air or vacuum and stresses the role of those
short-range interactions which occur at atomic contact. In foot-
note 46 Ninham hints at this aspect of short-range forces but does
not discuss it in very great detail.

References

[1] F. London, Zeit. Phys., 1930, 63, 245-279.

[2] W.H. Keesom, Phys. Zeit., 1921, 22, 129-141.

[3] P. Debye, Phys. Zeit., 1920, 21, 178-187.

[4] D. Tabor, "Solids, Liquids and Gases", Camb. Univ. Press, 1979, p. 19-21.

[5] H.B.G. Casimir and D. Polder, Phys. Rev., 1948, 73, 360-372.

[6] A.D. Buckingham, in "Intermolecular Interactions: From Diatomics to Polymers" (Ed. B. Pullman) Wiley, Chichester 1978, Chap. 1.

[7] R.G. Gordon and Y.S. Kim, J. Chem. Phys., 1972, 56, 3122-3133.

[8] M.J. Clugston, Advances in Physics, 1978, 27, 893-912.

[9] H.C. Hamaker, Physica, 1937, 4, 1058-1072.

[10] J.M. Corkhill, J.F. Goodman, C.P. Ogden and J.R. Tate, Proc. Roy. Soc. 1963, A273, 84-102.

[11] J. Requena, D.F. Billett and D.A. Haydon, Proc. Roy. Soc. 1975, A347, 141-159.

[12] C.J. Coakley and D. Tabor, J. Phys. D: Appl. Phys. 1978, 11, L77-L82.

[13] F.W. Fowkes, J. Colloid Interface Sci., 1968, 28, 493-

[14] J.F. Padday, and N.D. Uffindell, J. Phys. Chem. 1968, 72, 1407.

[15] J.N. Israelachvili and D. Tabor, Prog. Surface Membrane Sci., 1973 Ed. Matejevic, New York, Vol. 7, 1-55.

[16] J.N. Israelachvili, Quarterly Reviews of Biophysics, 1974, 6, 341-387.

[17] D. Langbein "Theory of van der Waals Attraction", Springer Tracts in Modern physics, Springer-Verlag, Berlin, 1974, 72.

[18] J. Mahanty and B.W. Ninham, "Dispersion Forces", Academic Press, New York, 1976.

[19] H. Margenau and N.R. Kestner "Theory of Intermolecular Forces", Pergamon, Oxford, 1971.

[20] V.A. Parsegian, IUPAC Commission I.6. Colloid and Surface Chemistry **4.1**, 1975, 27-72.

[21] J.N. Israelachvili, *J. Chem. Soc. Faraday Trans*. 1973, **68**, 1729-1738.

[22] R.M. Pashley and J.N. Israelachvili, *Colloids and Surf*., 1981, **2**, 169-187.

[23] D. Tabor and R.H.S. Winterton, *Proc. Roy. Soc.*, 1969, A**312**, 435-450.

[24] J.N. Israelachvili and D. Tabor, *Proc. Roy. Soc.*, 1972, A**331**, 19-38.

[25] C.H. Anderson and E.S. Sabinsky, *Phys. Rev. Letters*, 1970, **24**, 1049-1052.

[26] D. Tabor, *J. Colloid Interface Sci*., 1977, **58**, 2-13.

[27] D. Chan and P. Richmond, *Proc. Roy. Soc*., 1977, A**353**, 163-174.

[28] R.G. Horn and J.N. Israelachvili, *Chem. Phys. Letters* 1980, **71**, 192-4.

[29] R.G. Horn, J.N. Israelachvili and E. Perez, *J. Physique*, 1981, **42**, 23-36.

[30] B.V Derjagin, V.M. Muller and Yu.P. Toporov, *J. Colloid Interface Sci*., 1975, **53**, 314.

[31] J.N. Israelachvili, *J. Colloid Interface Sci*., 1980, **78**, 260-1.

[32] K.L. Johnson, K. Kendall and A.D. Roberts, *Proc. Roy. Soc*., 1971, A**324**, 301.

[33] V.M. Muller, V.S. Yushchenko and B.V. Derjagin, *J. Colloid Interface Sci*., 1980, **77**, 91-101.

[34] B.V. Derjagin, N.A. Korotova and V.P. Smilga, "Adhesion of Solids" Consultants Bureau, New York, 1978.

[35] S.S. Yoyutskii, "Autoadhesion and Adhesion of High Polymers", Wiley, New York, 1963.

[36] J.A. Greenwood and K.L. Johnson, *Phil. Mag*., 1981, **43**, 697-711.

[37] D. Maugis and M. Barquins, *J. Phys. D. Appl. Phys*., 1978, **11**, 1989-2023.

[38] J. Ferrante and J.R. Smith, *Surf. Sci*., 1973, **38**, 77-92.

[39] J.E. Inglesfield, *J. Phys. F: Metal Phys*., 1976, **6**, 687-701.

Addendum: In relation to retarded van der Waals forces, sophisticated theory shows that the loss in correlation occurs at distances greater than $\lambda/2\pi$ rather than λ but this does not change the type of argument used in deducing equation (3).

3
Fundamentals of Electrical Double Layers in Colloidal Systems

By J. Lyklema

LABORATORY FOR PHYSICAL AND COLLOID CHEMISTRY, AGRICULTURAL UNIVERSITY,
DE DREIJEN 6, 6703 BC, WAGENINGEN, THE NETHERLANDS

1. Introduction

Electrical double layers at phase boundaries play a dominant role in a range of colloidal phenomena, such as stability, electrokinetics, micelle formation, electrosorption, flotation, and polyelectrolyte properties. For particulate systems (sols, suspensions) the double layers consist of a charge on the particle, the *surface charge* σ_o, and an equal but opposite charge in the solution, the *countercharge*. For easy reference we shall use the term "particle" also for liquid drops (as in emulsions) or even for an air bubble (as occurs in froth flotation).

As surface charge and countercharge attract each other by Coulombic forces, a double layer would never form spontaneously if there were no other forces operative. One of these forces is of an entropical nature; it tends to spread ions equally over the available space and promotes the establishment of a diffuse double layer. On the other hand, there are "chemical" forces that can overcome Coulombic repulsion, leading to preferential accumulation at the interface of some types of ions in a mixture. Henceforth we shall designate all interactions that are neither Coulombic nor entropic as *specific*. They may be of a chemical or physical nature. By contrast, Coulombic forces are *generic*, i.e. the same for each species of the same valency: the binding energy is $zF\phi$, where ϕ is the electric potential, z is the valency of the ion with the sign included, and F is the Faraday constant.

The primary double layer parameters are *charge* and *potential*. Therefore, this review is mainly concerned with the description of both the origin and distribution of charges and potentials in double layers, and their interrelation. An important recurring matter is the question, what can be measured and by what technique. A derived property is the double layer *capacitance*, and this will also be included in the discussion.

Several double layer conditions and states are of practical relevance: isolated double layers versus double layers in interac-

tion (as for example giving rise to colloid stability), and double
layers in equilibrium (i.e. *relaxed* double layers) as opposed to
disequilibrated double layers (i.e. double layers that, due to an
outer force, are brought out of their equilibrium state and have
not yet had the time to return to this state). Nonequilibrium
double layers are of importance in the understanding of colloid
stability as well as electrokinetic, electrodekinetic, rheological and dielectric phenomena. Nonequilibrium and/or interacting
double layers can only be described if isolated relaxed double
layers are understood first. Therefore, this review which deals
with relaxed, isolated double layers can form the basis for attempts to analyse nonequilibrium and/or interacting double layers.

2. Some general features of electrical double layers

Electrical double layers are electroneutral entities. For our
purposes, this fact has at least two consequences.

First, it must always be possible to describe double layers
thermodynamically (i.e. operationally) as the outcome of adsorption of *electroneutral* species. The surface excess cr -deficit Γ_i
of a given electroneutral species i is experimentally accessible.
How the positive and negative charges are distributed in the
double layer cannot be established by adsorption measurements only.

Second, at any location x in the double layer one can always
say that the charge to the left of x, expressed per unit area, is
$\sigma(x)$; that to the right is $-\sigma(x)$. Then one can *by definition* call
these charges *the* surface charge and *the* countercharge, respectively. It follows that the notion "surface charge" is subject to
some element of definition, so that in actual situations it must
always be specified how σ_o is defined. One case where this need is
immediately obvious is that where adsorption studies are done in
conjunction with electrokinetic measurements. From the latter one
can obtain the *electrokinetic potential* ζ, from which an electrokinetic surface charge σ_{ek} can be computed. However, σ_{ek} is associated with the slipping plane which is known to differ from the
surface proper by a layer of a few molecular cross sections.
Hence, $\sigma_{ek} \neq \sigma_o$. Another example is that of double layers on
oxides. Here, σ_o is due to adsorption of H^+ and OH^- ions, but if
phosphate ions are also present in the solution, these ions sometimes chemisorb and contribute to the charge of the solid particle. It is then a matter of taste rather than of principle if the
charge contributed by the phosphate is included in the surface
charge. The two alternatives correspond to two differing locations
of the "surface".

Surface charge is an extensive parameter, but potentials are intensive. Surface charges are related to amounts of ions, whilst potentials are connected with the *electrical work* involved in bringing ions from one location (the reference state) to another. We shall choose the bulk of the liquid, at infinite distance from the surface, as the reference point.

The fact that double layers as a whole are electroneutral does not mean that the potential difference ϕ_o between a location near the centre of the particle and one in the bulk of the solution is zero. Henceforth, we shall call ϕ_o the *surface potential*. In fact, ϕ_o is an experimentally inaccessible quantity because there is no way by which one can split the total (electrochemical) work into an electrical and a "chemical" part. In the well-known formula for the electrochemical potential, the separation according to

$$\tilde{\mu}_i = \mu_i + z_i F \phi_o \tag{1}$$

does not take us further because both μ_i for single ionic species and ϕ_o are inaccessible. Only if one of the two is fixed by some agreement, the other is automatically defined. For instance, if $z_i F \phi_o$ is interpreted as a purely Coulombic term, as is usually done, all remaining terms are then collectively defined as μ_i. In passing, it is noted that in (1) all three terms are counted with respect to the bulk of the solution. Although ϕ_o is inaccessible, changes, $d\phi_o$, are under certain favourable conditions measurable.

Capacitances are derived quantities and are directly measurable on a number of systems. In principle they are a measure of the extent of the *screening* of the surface charge: the better σ_o is screened by the countercharge, the higher it can become at given ϕ_o. As the screening is related to the dielectric permittivity and to the extent of ion accumulation in the form of countercharge, capacitances are sensitive to the *structure* of the double layers.

Two different capacitances can be defined. The *integral capacitance*

$$K = \frac{\sigma_o}{\phi_o} \tag{2}$$

and the *differential capacitance*

$$C = \frac{d\sigma_o}{d\phi_o} \tag{3}$$

Differential capacitances are more commonly used for two reasons: (1) ϕ_o is inaccessible but $d\phi_o$ can be measurable, and (2) in

various experimental techniques the variation of σ_o due to a given
change of σ_o is measured (this applies in particular to bridge
measurements at polarized interfaces over which a given potential
can be applied; in that case the double layer is not relaxed with
respect to charge transfer).

The two capacitances are interrelated through the expressions:

$$K = \frac{1}{\phi_o} \int Cd\phi_o \qquad (4)$$

$$C = K + \phi_o \frac{dK}{d\phi_o} \qquad (5)$$

The integration in (4) has to start from some pre-established zero
point of charge, to be discussed in the following section.

3. Further discussion of charges and potentials

We shall now analyse the composition of double layers in some
more detail. By way of introduction, table 1 reviews some important processes through which surfaces can acquire a charge.

Table 1. Origins of charges at interfaces

	nature of interface	
	air/water, mercury/water, oil/water	solid/water
preferential adsorption of ions	+	+
dissociation of surface groups	−	+
isomorphic substitution	−	+
adsorption of polyelectrolytes	+	+
accumulation of electrons	+	+

Preferential adsorption of ions is the most common. In this
category can be included the adsorption of Ag^+ and I^- on silver
iodide particles, of H^+ and OH^- on insoluble oxides and of ionic
surfactants on many different types of particles. At the oil/water
interface, ionic surfactants can accumulate, but in their absence
there is often some tendency for anions to enrich preferentially
the oil phase, because of their (usually) higher polarizability.
For this reason, oil droplets in water tend to be negatively charged. At air/water interfaces anions are as a rule more easily dehydrated than cations and hence tend to enrich the air-side of
this interface, rendering it negative.

Dissociation of surface groups is a common feature with latices that can become charged by dissociation of, say, sulphate-, carboxyl- or amino-groups at the surface. The distinction between this behaviour and that of the first group is not sharp because, for instance, adsorption of an OH^- ion can not be experimentally distinguished from desorption of H^+.

Isomorphic substitution of ions is commonly found in clay minerals. Inside the crystal lattice of which the particles are formed Si^{4+} ions are sometimes replaced by Al^{3+} or other cations of lower valency, or Al^{3+} is replaced by Mg^{2+} whilst the overall crystal structure is retained. By this process the entire particle acquires a negative charge that is compensated by an excess of positive charge in the solution around the particle. Clay chemists express this countercharge in meq. 100 $gram^{-1}$ and call it the *cation exchange capacity* (c.e.c.). For our purposes, the charge per unit area acts as a surface charge; the fact that it is a bulk charge rather than a surface charge has no effect on the distribution in the solution side of the interface.

Adsorption of charged macromolecular substances is of great practical importance. Many biological polymers belong to this category. However, the theoretical understanding of polyelectrolyte adsorption is still in an embryonic stage.

Accumulation or depletion of electrons is the main charging mechanism of metal-solution interfaces in the absence of Faradaic currents (so-called polarized interfaces).

Because of the (sometimes inadvertently) omnipresence of small concentrations of electrolyte in any aqueous system it is likely that a dispersed particle will be charged. Actually, special conditions have to be created to render $\sigma_o = 0$. The most familiar of such conditions is to have a composition of the aqueous solution that is such that cations and anions adsorb to an equivalent amount. This situation is known as the *point of zero charge* (p.z.c.). For AgI it is a specific value of either the Ag^+ or the I^- concentration, usually expressed as $pAg^o = -\log c^o_{Ag^+}$, where the superscript ^{o} denotes the zero point condition. For this system pAg^o and pI^o are related through $pAg^o + pI^o = pL$, where L is the solubility product of AgI. By the same token, for oxides the p.z.c. is a certain value of pH, say, pH^o. The more acid the oxide, the lower the p.z.c.. For instance, with SiO_2, depending on details of pretreatment, $pH^o = 2 - 3$, but for haematite (α-Fe_2O_3), $pH^o = 8.5$. Silica and haematite are common components of soils. If these soils are slightly acidic, the first material is negatively charged, the second positive. Obviously this has important repercussions for the structure formation in soils because of the ensuing particle aggregation. It is also crucial for ion binding in these soils: for instance trace elements may become unavailable to plants.

What has been said for *aqueous* systems applies mutatis mutandis to *non-aqueous* ones. If the dielectric permittivity is very low, as in alkanes and many oils, electrolyte dissociation is suppressed and few ions or none at all are left in the system. Therefore, surface charges, if any, tend to be extremely low. Characteristically, there are only a few ionic charges per particle.

That σ_o is very low in such systems does by no means imply that ϕ_o is also low, because it is equally typical that the capacitances are very low: the screening power is extremely low. In fact, under certain conditions very high potentials can be generated due to the accumulation of relatively small amounts of charge. This occurs for instance if oil flows through pipelines; sparking has here sometimes given rise to fires. From a colloid stability viewpoint, it is important to realize that low σ_o values may suffice to induce resistance to aggregation.

With respect to their adsorption behaviour, it is customary in colloid science to distinguish between *potential determining* (p.d.) and *indifferent* ions. These terms have been coined by Verwey in the nineteenthirties. Taking the AgI system as the example, Ag^+ and I^- ions were identified as potential determining, because they are the sole species through which the solid could be charged and because they are at the same time lattice ions. Adsorption of such ions, therefore, did not lead to changes in the composition of the particles. Hence, the electrochemical potential $\tilde{\mu}^S_{Ag^+}$ of an Ag^+ ion in the solid simply equals $\mu^{oS}_{Ag^+} + F\phi^S$. As in the solution $\tilde{\mu}^L_{Ag^+} = \mu^{oL}_{Ag^+} + RT \ln c_{Ag^+} + F\phi^L$, it follows immediately that

$$F(\phi^S - \phi^L) = F\phi^o = \mu^{oL}_{Ag^+} - \mu^{oS}_{Ag^+} + RT \ln c_{Ag^+}$$

or

$$d\phi^o = \frac{RT}{F} d \ln c_{Ag^+} = - \frac{RT}{F} dpAg \tag{6}$$

which is *Nernst's law*. Choosing the p.z.c. as the reference point,

$$\phi^o - \phi^o (p.z.c.) = - \frac{RT}{F} (pAg - pAg^o) \tag{7}$$

Ions not obeying this law, e.g. K^+ and NO_3^- ions for AgI, were classified as *indifferent*.

Recently, the strict contradistinction between p.d. and other ions has met some criticism, because many systems do not behave in such an ideal fashion as Ag^+ and I^- ions do for AgI. For oxides, H^+ and OH^- are the charge generating ions, but as they are not constituent ions, their adsorption leads to compositional changes and Nernst's law does not apply to the surface potential, although it does to the potential difference between the two bulk phases (in this case these two potentials are different), as demonstrated by glass electrodes that behave like pH electrodes over many decades. Phosphate ions chemisorbing on oxides can hardly be called indifferent and neither can tenaciously adsorbing ionic surfactants.

Because of these considerations we shall henceforth refrain from using the notion "p.d. ion", except under ideal conditions. Ions contributing to σ_o will be called *charge determining* (c.d.) ions, and if necessary they will be specified, realizing that choosing them implies locating the surface (sec.2).

As a sequel to these considerations, it is now possible to write explicit expressions for σ_o. For the AgI case there are two options:

$$\sigma_o = F(\Gamma_{Ag^+} - \Gamma_{I^-}) \tag{8}$$

and

$$\sigma_o = F(\Gamma_{AgNO_3} - \Gamma_{KI}) \tag{9}$$

It is a matter of preference, but perhaps also a matter of principle, which one should be taken. The former expression emphasizes the ionic character and is useful if the double layer structure is to be analyzed in terms of ionic distributions, whereas the latter is more operational in that it expresses that the double layer as a whole is electroneutral, so that neither the adsorption of Ag^+ nor I^- can be measured without the accompanying adsorption of NO_3^- and K^+. However, also in the second case the interpretational (i.e. non-thermodynamic) step is made that $(\Gamma_{AgNO_3} - \Gamma_{KI})$ is interpreted as the charge sitting on the AgI surface. If the adsorption is elaborated thermodynamically using Gibbs' law, with (8) one needs *electrochemical* potentials of individual ions (i.e. inaccessible quantities) and an auxiliary electroneutrality condition. With (9) *chemical* potentials suffice. Obviously, because there is solid AgI present, the auxiliary condition $d\mu_{Ag^+} + d\mu_{I^-} = d\mu_{AgI}$ or $d\mu_{AgNO_3} + d\mu_{KI} = d\mu_{AgI} + d\mu KNO_3$ is required.

The AgI case is also suitable for a brief discussion of *negative adsorption*. From, say, a negative surface, anions are expelled so that they enrich the equilibrium solution, as compared with the situation in the absence of the charged colloidal particles. As double layer and bulk are electroneutral, this is experimentally observed as an increment in the neutral salt concentration. The Donnan equilibrium is a well-known consequence, and salt-sieving by compact charged sediments is an important application: if salt water percolates over such a sediment, the salt is expelled so that the efflux is salt-free. This is an important procedure to obtain drinking water from sea water.

Thermodynamically, there will be three surface excesses: Γ_{AgNO_3}, Γ_{KI} and Γ_{KNO_3}. The first two are positive, the last one is negative. The surface charge follows from (9). The countercharge consists of two parts: an excess Γ_{K^+} of K^+ ions, equal to $(\Gamma_{KI} + \Gamma_{KNO_3})$ (it is positive because on a negative surface $\Gamma_{KI} > \Gamma_{KNO_3}$) and a deficit of NO_3^- ions $\Gamma_{NO_3^-} = \Gamma_{AgNO_3} + \Gamma_{KNO_3}$, where $\Gamma_{NO_3^-} < 0$ because on a negative surface $|\Gamma_{KNO_3}| > |\Gamma_{AgNO_3}|$, Γ_{KNO_3} being negative.

Similar equations apply to other systems. For oxides,

$$\sigma_o = F(\Gamma_{HNO_3} - \Gamma_{KOH}) \tag{10}$$

If other chemisorbed species contribute to the charge, their surface excesses should be added to (10). In the case of surfactant adsorption, the molecule is as a rule not fully dissociated, i.e. for an adsorbate of sodium dodecyl sulphate, abbreviated as Na^+A^-, part of the Na^+ ions remains bound. By analogy with the above it is logical to define in this case also the surface charge as

$$\sigma_o = -F\Gamma_{A^-} \tag{11}$$

so that all the Na^+ are part of the countercharge, that is both the bound and the free fraction of Na^+. This is compatible with the conventional way of treating the double layer on AgI etc.: there all K^+ ions are counted as belonging to the countercharge, irrespective of whether or not they are strongly bound to the surface.

4. Information from electrokinetic measurements

At this instance it is expedient to pursue the difference between σ_o and σ_{ek} a bit further. From experience one is led to the

conclusion that if liquid moves tangentially with respect to a
solid surface a thin layer of the liquid, perhaps a few molecular
cross sections thick, remains immobile or at least much less mobile than the bulk liquid. Probably the fluidity changes smoothly
but rapidly over this layer, but for purpose of analysis it is
customary to replace this smooth fluidity curve by a step function
from zero to the bulk value at a given, but unknown, distance
d_{ek} from the surface. The layer in the liquid parallel to the surface at distance d_{ek} is called the *slipping plane*, the potential
$\phi(x = d_{ek})$ is identified as ζ and the charge, expressed per unit
area, σ_{ek}. Both ζ and σ_{ek} are experimentally accessible, but it
is not easy to assign a rigorous physical meaning to them, because
the slipping process is only poorly understood.

As has already been said, σ_{ek} and σ_o do not resemble each other
except under special conditions. Hence, electrokinetic measurements are unsuitable for the determination of the surface charge.
On the other hand, if σ_o and σ_{ek} can be measured simultaneously,
the comparison of these charges gives some information on the
charge *distribution* in the double layer. For many purposes, understanding colloid stability among them, the distribution of the
charge (in the sense of the compactness of the double layer) is at
least as crucial as its absolute value. Let us call the charge between surface and slipping plane σ_b, then because of the electroneutrality condition:

$$\sigma_o + \sigma_b + \sigma_{ek} = 0 \tag{12}$$

Note that unfortunately it is customary to give σ_{ek} the same sign
as ζ. If σ_o is < 0 and $\sigma_b \sim 0$, $\phi_o < 0$, $\zeta < 0$. This would render $\sigma_{ek} < 0$.
The idea is that σ_{ek} is counted as the charge on the particle side
of the slipping plane. From σ_o and σ_{ek}, the "electrokinetically
bound" charge can therefore be computed. Very generally speaking,
the following picture emerges from such experiments. $|\sigma_{ek}|$ seldom
exceeds $2 - 3$ μC cm^{-2}. For low σ_o, up to a few μC cm^{-2}, σ_b can be
low or even zero; then σ_o and σ_{ek} are similar or equal. If, however, σ_o is made to increase, all the additional surface charge is
compensated by co-adsorbing counterions in the immediate neighbourhood of the surface, without any effect on σ_{ek}. Typical examples
are oxides and ionic surfactant micelles, where σ_o can easily reach
values of several tens of μC cm^{-2}, so that values in excess of 90%
of this charge may be compensated by immobile or electrokinetically
bound counterions. A similar feature occurs with polyelectrolytes,

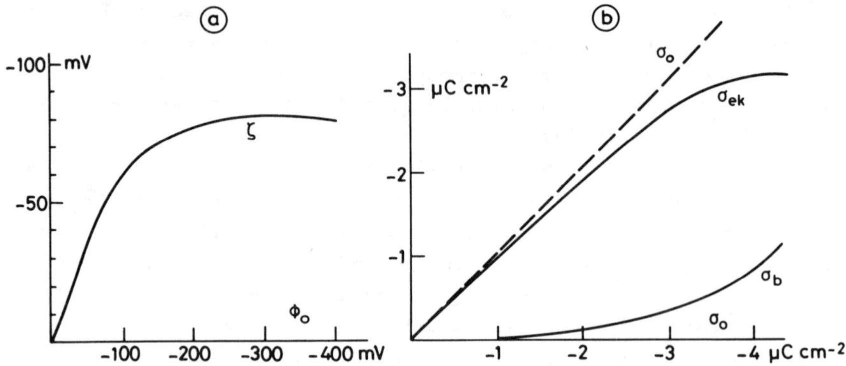

Figure 1. (a) Electrokinetic potential as a function of surface potential for AgI in 12×10^{-3} mol dm^{-3} KNO_3 (Troelstra et al., 1942); (b) distribution for this system of charge over electrokinetically mobile (ek) and bound (b) fractions.

where it is known as *counterion condensation*. Concomitant with this, if ζ is plotted as a function of ϕ_o, the initial part of the curve is linear, but then ζ levels off to some plateau value. Fig. 1 gives an example for a system where σ_b is relatively low.

The binding of ions in the electrokinetically immobile layer is partly due to the high potential in that layer, ϕ_s. The electrical binding energy is $z_i F \phi_s$. In addition, there can be other binding mechanisms (such as polarization, dipolar, quadrupolar, entropic and other contributions), collectively known as *specific adsorption* (s.a.). Due to s.a., σ_b is higher than it would be if no such specific effects were present. If s.a. is strong, σ_b may exceed σ_o; in this case σ_{ek} must have the same sign as σ_o to retain electroneutrality [*)]. The change of sign of σ_{ek} due to this so-called *superequivalent adsorption* of counterions is known as *charge reversal*. It should be noted that the reversal applies only to σ_{ek}, and not to σ_o; the latter quantity even *increases* somewhat due to induction.

If the conditions in the solution are such that $\zeta = \sigma_{ek} = 0$, the system is at its *isoelectric point* (i.e.p.). Just as with the p.z.c., for AgI the i.e.p. is a certain value of pAg and for oxides it is a certain pH value. In the absence of s.a., the p.z.c.

---

*) If σ_{ek} is counted as the charge on the particle side of the slipping plane, it bears a charge opposite to that of σ_o.

and i.e.p. coincide. However, if $\sigma_b \neq 0$ the two are different and
they shift even in different directions with increasing s.a..

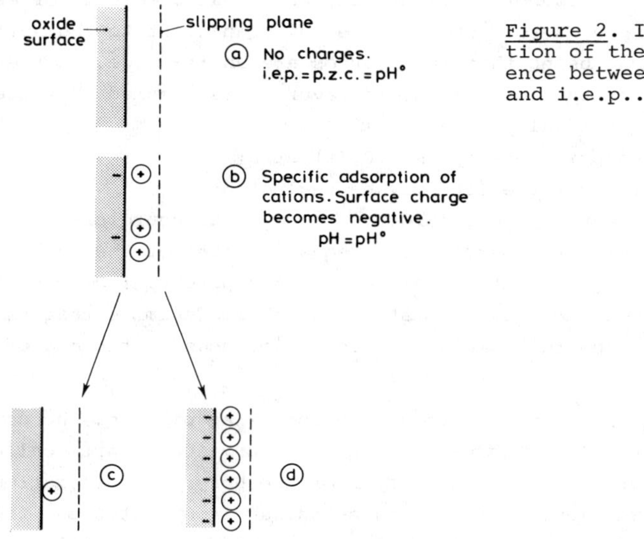

Figure 2. Illustration of the difference between p.z.c. and i.e.p..

   Figure 2 illustrates this, taking oxides as the example. In
fig. 2(a), $\sigma_o = \sigma_b = \sigma_{ek} = 0$, and the p.z.c. and i.e.p. coincide.
If now, at constant pH = pH°, some cations adsorb specifically
(fig. 2(b)), the surface acquires some negative charge due to induction; in this situation we are neither at the p.z.c. nor at the
i.e.p.. The new p.z.c (fig. 2(c)) is reached by compensating the
negative surface charges by proton adsorption, i.e. by lowering of
pH. However, the new i.e.p. (fig. 2(d)) is attained by making σ_o
more negative until $\sigma_o + \sigma_b = 0$; that is achieved by increasing
the pH.

   Simultaneous measurement of shifts in p.z.c. and i.e.p. is one
of the tests for specific adsorption.

5. Double layer theory

   Although the analytical and electrokinetic techniques discussed
so far do not exhaust the experimental procedures to obtain information on double layers, theoretical pictures must also be examined if further progress is to be made.

   One of the intrinsic problems is that the ion distribution in

the solution is governed by physical and chemical factors, and
these are in principle inseparable. For all practical purposes,
however, the assumption is usually made that chemical forces (that
is: s.a.) play a role only in the layer adjacent to the surface,
the argument being that such forces are short-range. In the remain-
der or outer part of the double layer it is assumed that the distribution is ideal in that:

1) the adsorption energy is $z_i F\phi(x)$ at x;
2) the volume of the ions may be neglected;
3) the solvent may be considered to be a structureless continuum
   with dielectric permittivity equal to that of the bulk;
4) the average potential $\phi(x)$ equals the potential of mean force.
   This is a subtlety of statistical thermodynamics that is beyond
   this review to explain (see the references at the end of this
   article).

Consequently, we arrive at the following picture. The double
layer consists of two parts: (a) an outer, ideal part, called the
diffuse part of the double layer or the *Gouy layer* (sometimes
Gouy-Chapman layer). It starts beyond a given distance d from the
surface and assumptions 1) - 4) apply. The plane parallel to the
surface at distance d is called the *outer Helmholtz plane* (OHp);
(b) an inner non-ideal part between the surface and d, where at
least one of assumptions 1) - 4) is involved; this part is called
the *compact* double layer part or *Stern layer*.

Since in the real system the Gouy layer is not rigorously ideal
any defects in the theory due to deviation from ideality in this
layer have also to be accounted for by the Stern layer. This un-
derlines that the distinction between a Gouy and a Stern layer is
somewhat artificial, although it appears to work well in practice.

Let us call the charge in the Stern layer σ_s and that in the
diffuse part σ_d. Then, because of electroneutrality,

$$\sigma_o + \sigma_s + \sigma_d = 0 \tag{13}$$

This equation is very similar to (12) and automatically the question arises by how much the slipping plane and the oHp differ.
Strictly speaking, the two can not be compared because both are
abstractions from a complex reality. However, for simple systems
like AgI, micelles, surfactant monolayers etc. there is considerable circumstantial evidence that for all practical purposes the
two may be identified, implying:

$$\sigma_s = \sigma_b \; ; \; \sigma_d = \sigma_{ek} \; ; \; \phi_d = \zeta \tag{14}$$

where ϕ_d is the potential of the diffuse part of the double layer, that is the potential at the oHp. For surfaces with gel-like coverages or "hairy" adsorbates, the situation is less well established.

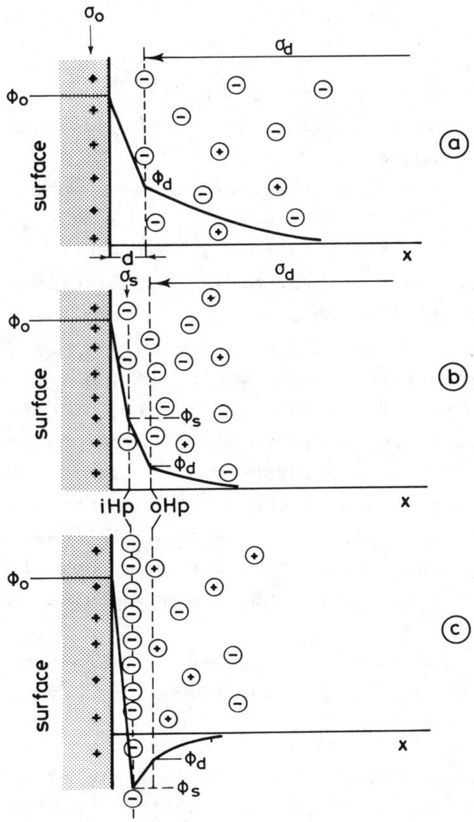

Figure 3. Gouy-Stern pictures of double layers.
(a) No specific adsorption.
(b) Some specific adsorption.
(c) Superequivalent specific adsorption.

Figure 3 gives pictorial representations of some double layer aspects of these; figure 3(a) is for the simplest case. In the absence of s.a. the entire countercharge is distributed diffusely. The Stern layer ($0 < x < d$) is charge-free and the charge balance is simply:

$$\sigma_o + \sigma_d = 0 \tag{15}$$

Situations like those in fig. 3(a) occur if the counterions are not easily dehydrated, so that if they retain their hydration

shells in the adsorbed state, d is directly related to the hydrated radius. The potential decay over the Stern layer is linear.
This is a consequence of Poisson's law, which for flat symmetry
can be written,

$$\frac{d^2\phi(x)}{dx^2} = - \frac{\rho(x)}{\varepsilon\varepsilon_o} \tag{16}$$

where $\rho(x)$ is the space charge density at x, ε the relative permittivity of the medium, and $\varepsilon_o = 8.854 \times 10^{-12}$ C V^{-1}m^{-1}. If $\rho = 0$
$d\phi/dx$ is constant. In other words: the electrical field strength
is constant.

In figure 3(b) there is some specific adsorption of anions inside the Stern layer. The plane through the centres of the specifically adsorbed ions is called the "*inner Helmholtz plane*" (iHp).
Usually, dehydration is a prerequisite for s.a.; only then can the
counterions come close enough to the surface. The angle between
the two straight portions in $\phi(x)$ at the iHp depends on σ_s, on the
dielectric permittivities in the two parts of the Stern layer, and
on the extent to which imaging of charges into the solid takes
place. For many practical cases one straight line from x =0 to
x = d is a satisfactory representation. The free energy of adsorption of ions at the iHp has a chemical term in addition to $z_i F\phi_s$.
In case 3(b) the value of ϕ_d is lower than in case 3(a).

Fig. 3(c) exemplifies superequivalent adsorption. Now $|\sigma_s| > |\sigma_o|$,
the signs of ϕ_d and ϕ_o are opposite and the signs of σ_d and σ_o are
the same.

The three pictures in fig. 3 are not the only possibilities.
Much more complex distributions may occur.

Let us digress for a moment into particle interaction. Double
layer pictures like those of fig. 3 are the basis for the interpretation of colloid stability phenomena. Stability is governed
by the overlap of double layers. In fact only the diffuse parts
of these contribute, because only these parts extend far into the
solution. The theory based on this premise is known as the Derya-
gin-Landau-Verwey-Overbeek (DLVO) theory. One illustrative equation of this theory relates the critical coagulation concentration
c_c (i.e. the concentration of indifferent electrolyte just enough
to destabilize a hydrophobic sol) to ϕ_d through:

$$c_c = \text{const.} \; \frac{[\tanh(ze\phi_d/4kT)]^4}{A^2 z^6} \tag{17}$$

Here A is the Hamaker constant, a measure for the attractive forces between the particles. As under the conditions where coagulation occurs ϕ_d is invariably so low that the hyperbolic tangent may be replaced by the first (linear) term of its series expansion,

$$c_c = \text{const'} \ \frac{\phi_d^4}{A^2 z^2} \tag{18}$$

it is clear that c_c is very sensitive to ϕ_d. Sol stability is known to decrease drastically with increasing z (the so-called rule of Schulze and Hardy). With (18) this can now be understood because, in addition to the z^2 in the denominator, there is a tendency for s.a. to increase with increasing z, that is: a trend of decreasing ϕ_d. Lyotropic sequences due to small differences with respect to σ_s are also reflected in c_c through different values of ϕ_d.

We recall that, because of arguments given before, for many systems the substitution $\phi_d \rightarrow \zeta$ is justified.

Pictures like those of fig. 3 point to different degrees of screening, depending on the extent of s.a.. A consideration of capacitances illustrates this. For situation (a), if the differential Stern capacitance C_s is defined as

$$C_s = \frac{\partial(\phi_o - \phi_d)}{\partial\sigma_o} \tag{19}$$

it follows immediately that

$$\frac{1}{C} = \frac{1}{C_s} + \frac{1}{C_d} \tag{20}$$

which is the familiar expression for two capacitors in series. The important consequence is that the overall capacitance is dominated by the *smaller* of the two. As it follows from Gouy theory that C_d increases rapidly with c_{salt} and ϕ_d, this implies that, except at low ϕ_d and c_c, the double layer capacitance is mainly determined by the properties of the Stern layer.

If there is s.a. in the Stern layer, instead of (20) we have:

$$\frac{1}{C} = \frac{1}{C_s} + \frac{1}{C_d} \ (1 + \frac{\partial\sigma_o}{\partial\sigma_s}) \tag{21}$$

The additional term accounts for the increase of σ_o (at fixed ϕ_o), due to induction. As $\partial\sigma_o/\partial\sigma_s$ is always smaller than unity, the tendency of s.a. is to increase C.

By way of illustration we shall now give a selection of important double layer formulas. For derivations and other equations, see the bibliography at the conclusion of this chapter.

For a flat double layer the Gouy theory can be elaborated analytically at any ϕ_d. The potential distribution in a flat diffuse double layer is obtainable from (16) by double integration after substitution of Boltzmann's law for the ionic concentrations. For a symmetrical (z-z) valent electrolyte it obeys

$$\tanh\left(\frac{ze\phi(x)}{4kT}\right) = \tanh\left(\frac{ze\phi_d}{4kT}\right)e^{-\kappa(x-d)} \tag{22}$$

where κ is the reciprocal Debye length, given by

$$\kappa = \left(\frac{2e^2nz^2}{\varepsilon\varepsilon_o kT}\right)^{\frac{1}{2}} \tag{23}$$

where n is the number of cations or anions per unit volume. At room temperature, in aqueous solution $\kappa \sim x\sqrt{10^{15}\ c}\ \mathrm{cm}^{-1}$, if c in mol dm^{-3}. The low potential approximation of (22) is:

$$\phi(x) = \phi_d e^{-\kappa(x-d)} \tag{24}$$

indicating that ϕ_d decays to ϕ_d/e over a distance κ^{-1}. This is why κ^{-1} is often called the diffuse double layer thickness.

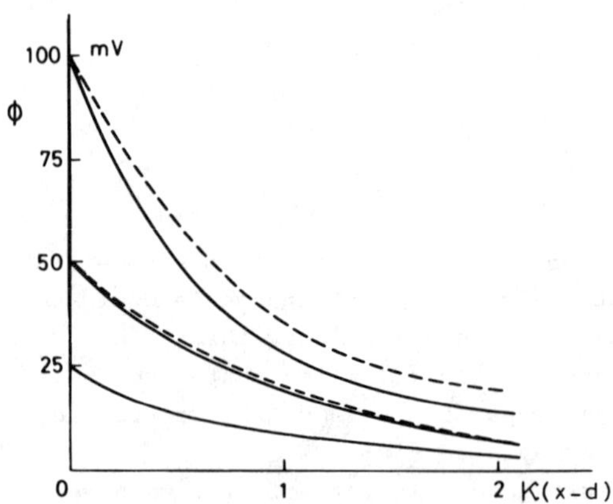

Figure 4. Potential distribution in a flat diffuse double layer. Drawn curves: eq.(22); dashed curves: eq.(24). z = 1, T = 25 °C.

Figure 4 illustrates this behaviour. Approximation (24) appears to be satisfactory up to about 50 mV. Plotting $\phi(x)$ as a function of the dimensionless parameter $\kappa(x-d)$ enables "scaling" in the sense that curves at different concentrations are obtainable by expanding or compressing the abscissa. It is a well-known feature that the potential decays more steeply with increasing κ.

The corresponding surface charge is given by

$$\sigma_d = (8\varepsilon\varepsilon_o nkT)^{\frac{1}{2}} \sinh(\frac{ze\phi_d}{2kT}) \tag{25}$$

corresponding to $11.72 \sqrt{c} \sinh(ze\phi_d/2kT)$ μC cm^{-2} for aqueous solutions at $25\ ^oC$ if c in mol dm^{-3}. The hyperbolic sine leads to a strong increase of charge with ϕ_d and c_{salt} as figure 5 illustrates. Differentiation of (25) with respect to ϕ_d yields C_d (see eq. (3)) which also increases progressively with c_{salt} and ϕ_d, so that in (20) C_d^{-1} rapidly becomes vanishingly small.

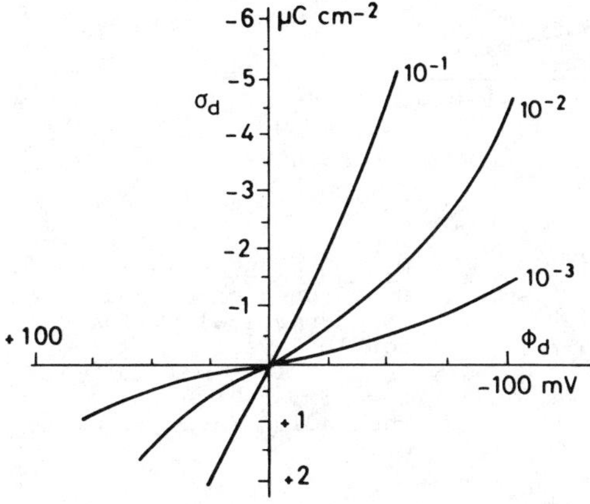

Figure 5. Charge in a flat diffuse double layer at various (1-1) electrolyte concentrations. T = 25 ^oC.

For spherical double layers no analytical solutions are available except at low ϕ_d. Because of the divergence of the lines of force a larger part of the countercharge is accomodated at relatively low potentials and this renders the low potential solution better than in the case of planar symmetry. The resulting equation reads

$$\phi(x) = \phi_d(\frac{a+d}{r}) e^{-\kappa[r-(a+d)]} \tag{26}$$

where a is the particle radius and r the distance to the centre of the particle. Figure 6 gives examples of distributions, showing the characteristic feature that the decay is the steeper, the stronger the curvature. For $q \to \infty$ the flat double layer case is recovered.

Corresponding plots for the surface charge are given in figure 7. In agreement with the trends observed above, σ_d increases very rapidly with ϕ_d, the more so the stronger the curvature which is clearly shown in this diagram.

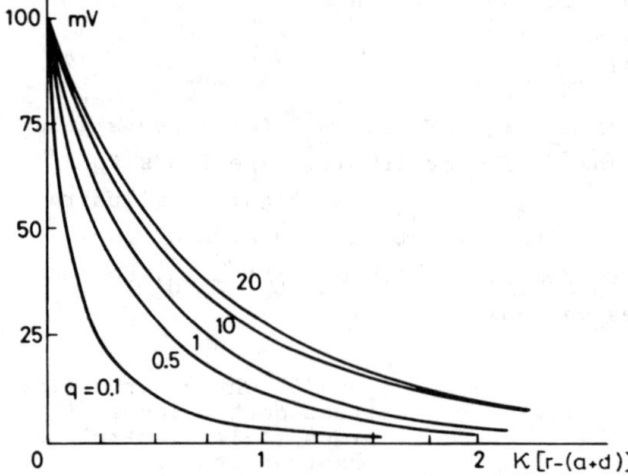

Figure 6. Exact solutions of the potential drop in a spherical diffuse double layer.
$\phi_d = 100$ mV; $z = 1$; $T = 25\ ^oC$.
The radius of particle and Stern layer follows from q by dividing by 3.2866 $\sqrt{c_{salt}}$, with c_{salt} in mol dm^{-3}.

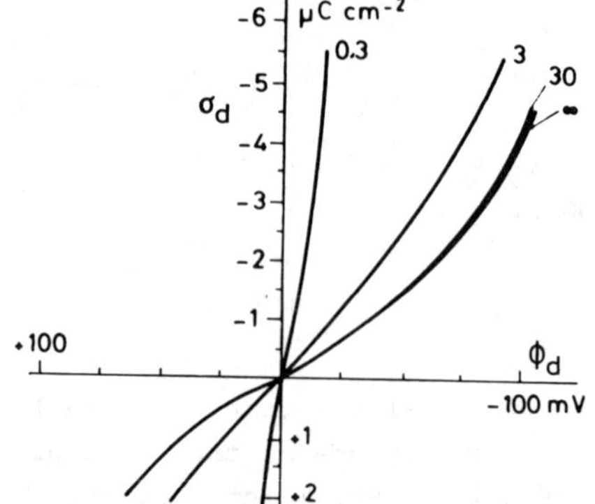

Figure 7. Surface charge as a function of potential in a spherical diffuse double layer.
$c_{salt} = 10^{-2}$ mol dm^{-3}; $z = 1$; $T = 25\ ^oC$.
The radius (a+d) in nm is given.

The treatment of the non-diffuse part of the double layer is mostly rather complex and general equations cannot be given. Usually, its composition is better inferred from a comparison between σ_o and σ_d (or, for that matter, between ϕ_o and ζ or ϕ_d) than computed by some model. The following equation has a reasonably wide applicability if s.a. is not superequivalent:

$$\frac{\theta}{1-\theta} = \frac{c_i}{55.5} \exp. - [\frac{z_i e \phi_s + \Phi_i}{kT}] \tag{27}$$

where

$$\theta = \sigma_s / \sigma_o \qquad (0 \leq \theta \leq 1) \tag{28}$$

and Φ_i stands for the specific adsorption free energy, and ϕ_s is the potential at the iHp. Eq.(27) is an extension of the familiar Langmuir equation, in which lateral interaction is accounted for in the so-called Bragg-Williams approximation. Using (27) remains intrinsically difficult because there is no a priori procedure to measure ϕ_s.

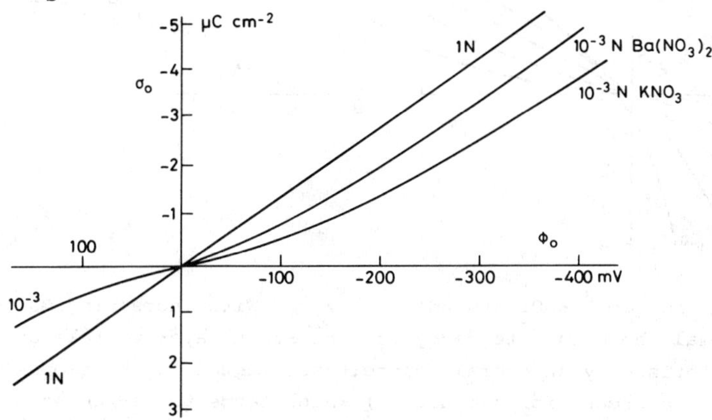

Figure 8. Surface charge in a Gouy-Stern layer in dilute and concentrated KNO$_3$ and Ba(NO$_3$)$_2$. The Stern layer is charge-free and C_s = constant = 15 µF cm^{-2}.

In conclusion of this discussion, we shall reproduce a few experimental $\sigma_o(\phi_o)$ curves.

Figure 8 is computed on the basis of (20), assuming C_s to be constant and equal to 15 µF cm^{-2}, and calculating C_d from Gouy theory. If this diagram is compared with figure 5 it becomes obvious that the introduction of a charge-free Stern layer has a profound effect on the overall behaviour. In 1 N electrolyte the diffuse part has entirely vanished, but in 10^{-3} N solution, especially around the p.z.c., some similarity with the purely diffuse behaviour can still be observed. Because of their higher valency, Ba^{2+} ions are more effective at screening of charge than are K$^+$ ions, so that on the negative side $\sigma_o(\mathrm{Ba(NO_3)_2}) > \sigma_o(\mathrm{KNO_3})$. On the positive side, however, the NO$_3^-$ is the counterion and there are no differences between KNO$_3$ and Ba(NO$_3$)$_2$ solutions of the same normality.

Figure 8 bridges the gap to experimental systems. Figure 9 gives classical curves for AgI. Here, σ_o is plotted as a function of pAg, which is related to ϕ_o through (7). Again, the hyperbolic sine functionality demanded by Gouy theory (25) is only observed

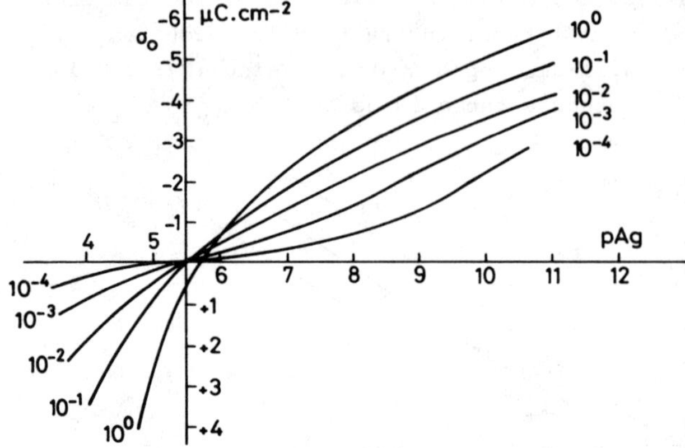

Figure 9. Surface charge on silver iodide suspensions in solutions of various concentrations of KNO_3. T = 25 °C.

in 10^{-3} mol dm^{-3} KNO_3 around the p.z.c.. With increasing c_{salt} and potential there is a tendency for the Stern layer to take over, characterized by an overall capacitance (equal to the slopes of the curves except for a constant) which tends to be the same for all curves at sufficiently negative ϕ_o (i.e. to the right in the figure).

The 10^0 curve is not linear, as in the calculated example of figure 8. In other words, C_s is a function of σ_o or ϕ_o. The 10^0 curve does not intersect the other curves at the p.z.c., but slightly to the right of it. In view of the discussion in connection with figure 2, this means that in this region there must be a slight s.a. of anions, i.e. NO_3^-.

It appears that on AgI cations adsorb also specifically, but they do this only at sufficiently high c_{salt} and ϕ_o. Figure 10 gives σ_o(pAg) curves in 10^{-1} mol dm^{-3} nitrates. Apparently, Rb^+ ions accumulate more strongly in the Stern layer than Li^+ ions do, because in (27) $\Phi_{Rb^+} > \Phi_{Li^+}$. This difference in adsorbability leads also to a lower ϕ_d in the case of Rb^+ and hence to a lower coagulation concentration (18). Surface charge curves and lyotropic sequences in the stability are therefore complementary pieces of information.

Silver iodide is one of the classical model systems of colloid science and studied in great detail. It appears that the substance is more or less typical for nonporous solids. It is gratifying that σ_o(ϕ_o) curves on AgI and Hg are also very similar. First, this supports the correctness of the measurements which are very different, for example the following differing techniques have

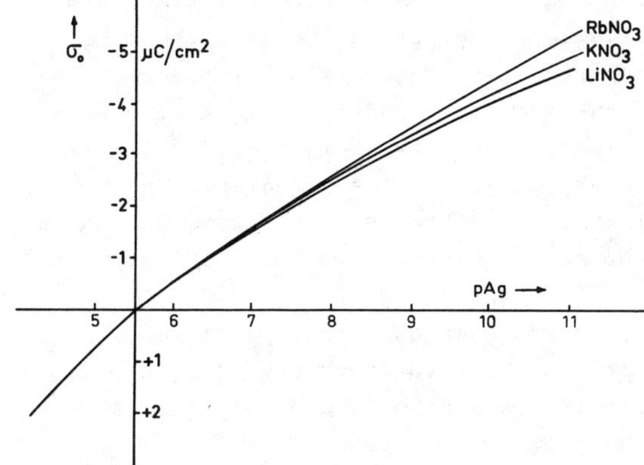

Figure 10. Lyotropic effects in the double layer charge on silver iodide.

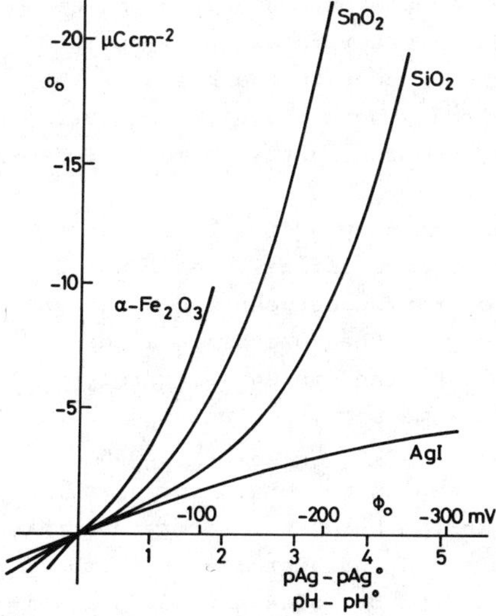

Figure 11. Surface charge on a number of insoluble oxides, with that on silver iodide included for comparison purposes.
Electrolyte: 10^{-1} KNO$_3$.
T = 25 °C.

been used: capacitance bridge or electrocapillary curves on Hg, direct measurement of σ_o, using (8) or (9) on AgI. Second, it indicates that the solution side controls the double layer properties. In the third place, it means also that various procedures of analysis that have been elaborated on Hg in great detail are also applicable to AgI and similar substances.

A group of solids that follows a distinctly different pattern is that of insoluble oxides, of which figure 11 gives an example. Such materials have in common that σ_o(pH) curves are convex with the pH axis and that σ_o can become very high, several times exceeding the corresponding value on AgI or Hg. Several theories have been forwarded to account for this behaviour. These theories must at least account for the fact that under a number of conditions σ_o can exceed the value dictated by (10) if all available surface groups are fully charged. It is likely that part of σ_o is not a surface charge in the strict sense of the word but has a certain depth, i.e. the surface is "porous" for c.d. and counterions.

Oxide double layers may be representative for particles with porous surfaces, whether or not these surfaces are of a gel-like nature. Perhaps such double layers are not too dissimilar from double layers around coiled polyelectrolytes.

6. Application to colloid stability

Understanding the composition of double layers around colloid particles is the first step towards an interpretation of colloid stability. By way of example, in the discussion the basis of the so-called Schulze-Hardy rule has already been given in connection with formulas (17) and (18). Let us now consider in some more detail particle interaction.

When two particles approach each other by diffusion or shear, they start to influence each other electrostatically as soon as their double layers overlap. The separation between the surfaces for this overlap to occur is the larger, the more diffuse the double layer is, that is: the lower the indifferent electrolyte concentration. For particles of the same sign, the ensuing interaction is always repulsive. If the particles come very close (and this is possible if the electrolyte concentration is increased), the omnipresent Van der Waals attraction between them may overtake the repulsion and force the particles to aggregate. This is the general principle of DLVO-theory, and it provides already a qualitative interpretation of the effect of electrolytes.

For a quantitative interpretation a number of points have to be taken into consideration. One of these points concerns the double layer structure: as the Stern layer is only a few molecular diameters thick, it plays no direct role in interaction, but its indirect role in dictating the value of ϕ_d is enormous. In DLVO theory the double layers are considered as if they were purely diffuse, but in reality the theory applies to the diffuse *parts* of the two interacting double layers, and the surface potential ϕ_o in DLVO-theory should rather be replaced by ϕ_d which, in turn, may be substituted by ζ for nonporous substances.

The next important step is to consider the *dynamics* of the interaction. Much depends on the time scale of particle interaction by diffusion relative to that of the adjustments of the double layers to the overlap. If the rate of approach is relatively fast, the double layers have little or no time to adjust, but if it is slow, the double layers are continually at equilibrium. This is in fact the path chosen in the DLVO theory; it is based on the consideration that diffuse double layers relax extremely fast. If this situation does prevail, the repulsion energy $V_R(h)$ can be computed by equilibrium thermodynamics as the isothermal reversible work to bring the particles from infinity to a distance h. The expression for $V_R(h)$ for interaction between flat double layers in a symmetrical (z-z) valent electrolyte reads:

$$V_R(h) = \frac{64\,nkT}{\kappa}\,[\tanh(\frac{ze\phi_d}{4kT})]^2\,e^{-2\kappa h} \tag{29}$$

Here the symbols have the same meanings as in (22), (23) and (25). The hyperbolic tangent occurs in (29) because of the hyperbolic tangent shape of the potential decay in a flat diffuse double layer (22). For spherical particles an equation similar to (29 holds.

In addition to the electrolyte concentration occurring several times in (29), it is characteristic that (29) contains ϕ_d rather than σ_d. The underlying idea is that because of equilibrium during encounter ϕ_d should remain constant, since it is determined by the adsorption of the various charge-determining ions, the chemical potentials of which are fixed by their chemical potentials in bulk (in the DLVO-theory this is straightforward because there ϕ_d is replaced by ϕ_o, which is related to the concentration of p.d. ions according to (7)). This type of interaction is known as *"interaction at constant potential"*. The alternative picture is *"interaction at constant charge"*, which is expected if the interaction proceeds so fast that no ionic species can adsorb nor desorb, In the latter case ϕ_o (or ϕ_d) is not constant but rises during the overlap. There is not a great difference in V_R for the two types of interaction, but the "constant charge" case tends to give somewhat higher repulsion energies.

Resolving the question which of the two mechanisms is operative in cases with model colloids is one of the challenging problems of modern stability research. For one thing, independent measurements of the dynamics of the various steps of charge transfer are needed. Recent experiments with silver iodide sols suggest that, in reality rather, a situation prevails intermediate between the extremes of "constant charge" and "constant potential" in that interaction takes place at constant potential and constant *total* surface charge, but at varying surface charge density distribution.

7. Summary

This contribution reviews a number of important properties of electrical double layers, such as charge, potential, capacitance, zero points of charge, isoelectric points, and their various relationships. Some basic equations of double layer theory are given and illustrated. Comparison is sought with experimental data, and the relevance of this work for electrokinetics and colloid stability is briefly elaborated.

8. Double layer books and reviews

Comprehensive Treatise of Electrochemistry, Vol. 1 "The Electrical Double Layer", J.O'M. Bockris, B.E. Conway, E. Yeager, eds., Plenum Press, New York-London, 1980.

M.J. Sparnaaij, "The Electrical Double Layer", Pergamon Press, Oxford-New York-Toronto-Sydney-Braunschweig, 1972.

G.R. Wiese, R.O. James, D.E. Yates, T.W. Healy, "Electrochemistry of the colloid-water interface", in "International Review of Science", Phys.Chem.Ser. Two, Vol. 6, A.D. Buckingham, J.O'M. Bockris, Consulting eds., Butterworths, London, 1976.

B.H. Bijsterbosch, J. Lyklema, "Interfacial Electrochemistry of Silver Iodide", Advan.Colloid Interface Sci., 1978, 9, 147.

D.C. Grahame, Chem.Revs., 1947, 41, 441.

C.A. Barlow, "The electrical double layer" in "Physical Chemistry, an Advanced Treatise", H. Eyring, D. Henderson, W. Jost, eds., Acad. Press, New York-London, 1970, Vol.IXA, Ch. 2.

4
Adsorption from Solution

By D. H. Everett

DEPARTMENT OF PHYSICAL CHEMISTRY, SCHOOL OF CHEMISTRY, UNIVERSITY OF BRISTOL,
CANTOCK'S CLOSE, BRISTOL BS8 1TS, U.K.

Introduction

In colloid science a major area of research is concerned
with the forces between particles of a dispersed phase which
determine not only whether a dispersion under given conditions is
stable but also many other properties of the system. These
interactions depend on the nature and the detailed structure of
the medium between the particles and this influence becomes
particularly significant when one or more of the molecular species
in the medium is preferentially adsorbed at the particle surfaces.

There are two important special cases. In the first the
medium contains ions and the surfaces are electrically charged:
these electrical double layer effects are discussed by Lyklema[1].
A second important group of phenomena arises when the adsorbed
species are polymeric, and these are dealt with by Napper[2]. In
both instances the forces between the particles depend on the
behaviour of the adsorbed layers as the particles are brought
together[3]. Thus an understanding of colloidal stability
depends on the availability of adequate theories of adsorption
from solution and of the structure and behaviour of adsorbed
layers.

Furthermore adsorption from solution plays an important role
in many other colloidal phenomena such as adhesion, wetting,
liquid/liquid displacement, detergency and lubrication. Other
important applications of adsorption include its use in a wide
range of purification processes and in liquid chromatography.

The object of this chapter is to outline some of the more
important themes in the study of adsorption from solution and to
indicate briefly their importance. Discussion will be limited
to relatively simple systems since, although in practice much
more complex systems are encountered, the subject has not yet

reached the point at which theoretical discussion of them is worthwhile.

Definitions and fundamental equations

The fundamental molecular situation leading to the concept of adsorption is that in which the local concentration of molecules in the neighbourhood of the surface or interface differs from that in the adjacent bulk phases (Figure 1).

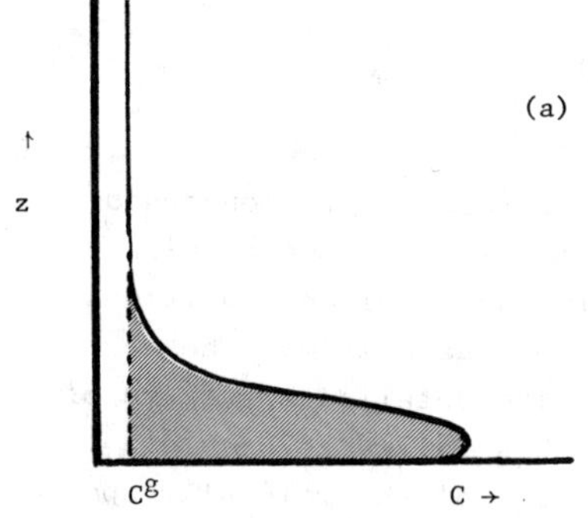

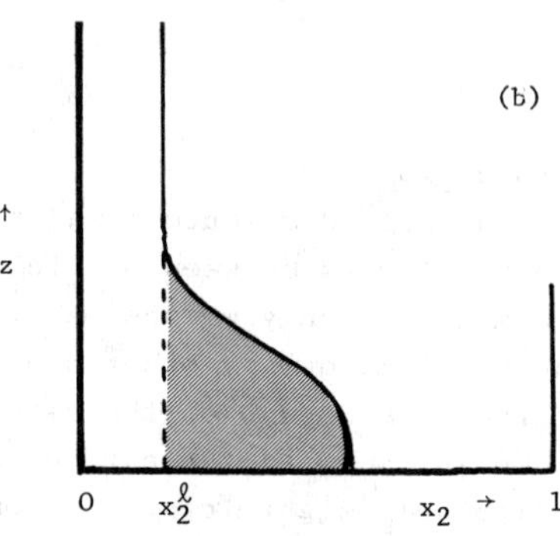

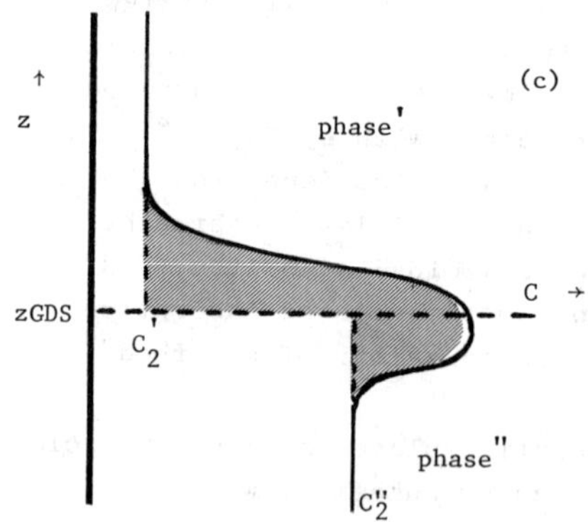

Figure 1: Concentration profiles in the vicinity of an interface.

(a) solid/single gas

(b) solid/liquid mixture

(c) fluid/fluid

z is the direction normal to the surface.

In the case of a single gas in contact with a solid, where the concentration of gas molecules close to the surface is higher than in the bulk gas (and is zero within the solid), the number of molecules present is greater than that which would be present if the bulk gas concentration were maintained up to the surface: there is, in effect, a <u>surface excess of molecules</u> defined by the shaded area in figure 1(a).

The situation is somewhat different for multicomponent systems exhibiting liquid / vapour, liquid/liquid or liquid/solid interfaces. Here if, in the surface layer, the concentration of one kind of molecule is greater than in the adjacent bulk liquid, then because of the essentially close-packed nature of the liquid, the concentration of at least one of the other components must be smaller, (figures 1(b) and 1(c)). In these cases the definition and measurement of adsorption(Γ) or surface excess, presents special problems which are, however, different in each case. Thus for the liquid/vapour and liquid/liquid interfaces direct experimental measurement of adsorption is difficult, except where insoluble films are formed at the inter-face. A precision of better than a few percent is difficult to achieve, in contrast to the high precision with which adsorption can be measured at both solid/gas and solid/liquid mixture interfaces. This asymmetry is further illustrated by the fact that whereas surface tension (σ) has a clear operational meaning for fluid / fluid interfaces, there is no simple direct and unambiguous experimental method of measuring the surface tension of a solid in contact with a gas or liquid. The importance of thermodynamics is that it provides, through the Gibbs adsorption isotherm (see below), a link between surface tension and adsorption so that, in each case, the quantity which is not readily accessible experimentally can be calculated from appropriately designed measurements of the other, together with information on the chemical potentials (μ) of the components in the bulk phases. The situation can thus be summarised as follows

	Fluid/Fluid (ℓ/g or ℓ/ℓ)	Fluid/Solid (ℓ/s or g/s)
Γ	Accurate measurement very difficult and usually not possible	Can be measured accurately
σ	Can be measured accurately	σ has no direct experimental (mechanical)meaning
	$\Gamma(x_i)$ can be calculated from knowledge of $\sigma(x_i)$	$\sigma(x_i)$ or $\sigma(p)$ can be calculated from $\Gamma(x_i)$ or $\Gamma(p)$

We shall now limit attention to the solid/liquid interface, and discuss in detail the measurement of adsorption from solution, the means by which thermodynamic quantities can be calculated from such measurements, and the molecular interpretation which can be put on such results.

Adsorption from Solution

Adsorption from a liquid mixture by a solid can be defined[4,5] precisely in operational terms by considering an experiment in which a given mass (m) of solid is brought into contact with an amount n^o of a liquid mixture (for convenience we consider here only a binary mixture) of mole fraction x_2^o: in general at equilibrium the composition of the bulk phase will have changed to x_2^ℓ. If the whole of the liquid phase were at this mole fraction the system would contain an amount $n^o x_2^\ell$ of 2: in fact it contains $n^o x_2^o$. The <u>surface excess amount of component 2</u> can thus be defined as

$$n_2^{\sigma(n)} = n^o(x_2^o - x_2^\ell) = n^o \Delta x_2^\ell. \tag{1}$$

The adsorption by unit mass of solid, the <u>specific surface excess</u>, is

$$n_2^{\sigma(n)}/m = n^o \Delta x_2^\ell/m, \tag{2}$$

and is the preferred form in which experimental data should be summarised.[*] If the <u>specific surface area</u> of the solid (a_s) is known, the <u>areal surface excess</u>, $\Gamma_2^{(n)}$, can be calculated:

$$\Gamma_2^{(n)} = n^o \Delta x_2^\ell/ma_s. \tag{3}$$

It also follows that since $x_1 + x_2 = 1$, $\Delta x_1^\ell = -\Delta x_2^\ell$,

$$\Gamma_1^{(n)} = -\Gamma_2^{(n)}; \tag{4}$$

---

[*]For some purposes, especially in dilute solution, it may be more convenient to measure the related quantity

$$n_2^{\sigma(v)}/m = V^o \Delta c_2^\ell/m$$

where V^o is the volume of solution and Δc_2^ℓ the change in concentration.

this relation stresses the competitive nature of the adsorption.

The definition of adsorption introduced by Gibbs[6] is at first sight more abstract. It is convenient to express the concept first in general terms and then apply it to the special case of solid/liquid interfaces. In figure 1(c) we choose a particular (but arbitrary) surface to define the interface (the so-called Gibbs dividing surface, GDS) between phases ' and " and calculate the difference, n_2^{σ}, between the amount of a given component actually present (n_2) and the amount which would be present in a system of the same volume if the bulk concentrations (c_2' and c_2'' respectively) had remained constant up to the chosen GDS which defines the volumes V' and V" of the two phases:

$$n_2^{\sigma} = n_2 - c_2'\, V' - c_2''\, V''; \tag{5}$$

this is represented by the shaded area in the diagram.

If we consider unit surface area of interface and columns of the phases ' and " out to z' and z" then the areal excess Γ_2 is given by

$$\Gamma_2 = \frac{n_2}{A} - c_2'\,(z' - z_{GDS}) - c_2''\,(z_{GDS} - z''). \tag{6}$$

It is clear from this equation and from simple geometrical considerations that if the location of the GDS is moved by δz then the value to be ascribed to Γ_2 changes by $\delta\Gamma_2$ where

$$\delta\Gamma_2 = (c_2'' - c_2')\delta z = \Delta c_2\,\delta z. \tag{7}$$

Γ_2 defined by equation (6) cannot therefore be employed to characterise the system unless some unambiguous means of fixing z_{GDS} can be found. Figure 2 shows diagrammatically the way in which Γ_2 and Γ_1 of a binary mixture vary with the location, z_{GDS} chosen for the GDS. One way of choosing the GDS is to locate it at $z^{(1)}$ where the adsorption of 1 is zero; the adsorption of 2 corresponding to this choice is written $\Gamma_2^{(1)}$ and called the <u>relative adsorption of 2 with respect to 1</u>. From the geometry of the diagram it is readily shown that $\Gamma_2^{(1)}$ is related to Γ_1 and Γ_2 defined relative to an arbitrary GDS by

$$\Gamma_2^{(1)} = \Gamma_2 - \Gamma_1 \frac{\Delta c_2}{\Delta c_1}; \tag{8}$$

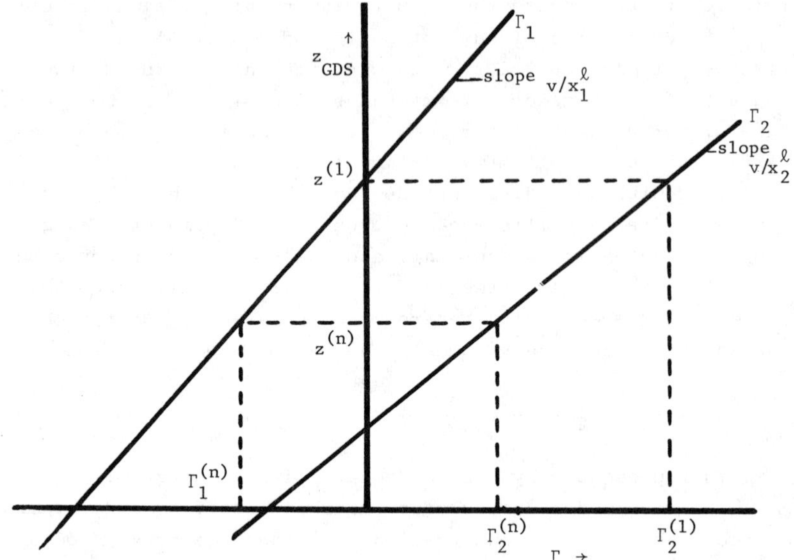

Figure 2: Dependence of adsorption Γ on location z_{GDS} chosen for the GDS $z^{(1)}$ and $z^{(n)}$ are the locations which define $\Gamma_2^{(1)}$ and $\Gamma_2^{(n)}$ respectively.

when $\Gamma_1 = 0$, $\Gamma_2 = \Gamma_2^{(1)}$. It may, furthermore, be shown(7) that $\Gamma_2^{(1)}$ has a well-defined experimental meaning:

$$\Gamma_2^{(1)} = \frac{1}{A}\left[(n_2 - Vc_2') - (n_1 - Vc_1')\frac{\Delta c_2}{\Delta c_1}\right] \tag{9}$$

where V is total volume of the system and n_1 and n_2 are the total amounts of 1 and 2. All quantities on the right hand side are in principle measurable. However, unless V is very small, the two terms in the square brackets are almost equal and their difference is very sensitive to small experimental errors.

The experimental quantity $\Gamma_2^{(n)}$ (sometimes called the reduced adsorption) is seen to correspond to choosing the GDS such that $\Gamma_1^{(n)} = -\Gamma_2^{(n)}$ i.e. so that the total adsorption is zero.

In the particular case of the solid/liquid interface the relation between $\Gamma_2^{(n)}$ and $\Gamma_2^{(1)}$ is readily obtained since here (taking the phases ' and " as s and ℓ and $c_1^s, c_2^s = 0$),

$$\Gamma_2^{(1)} = \Gamma_2^{(n)} - \Gamma_1^{(n)} \left[\frac{c_2^{\ell}}{c_1^{\ell}} \right] = \Gamma_2^{(n)} \left[1 + \frac{c_2^{\ell}}{c_1^{\ell}} \right]. \qquad (10)$$

Since $c_2^{\ell}/c_1^{\ell} = x_2^{\ell}/x_1^{\ell}$ this reduces to

$$\Gamma_2^{(1)} = \Gamma_2^{(n)}/x_1^{\ell}. \qquad (11)$$

The importance of this equation is that $\Gamma_2^{(1)}$, the quantity which appears most directly in thermodynamic equations, can thus be calculated immediately from the experimental quantity $\Gamma_2^{(n)}$. Furthermore, we see that $\Gamma_2^{(n)}$ corresponds to choosing the solid surface as the GDS.

Two comments are important. First, the Gibbs method is nothing more than a book-keeping operation which ensures overall stoichiometric consistency. It should not be called the Gibbs model since this is sometimes, erroneously, taken to imply that molecules of finite size comprising the surface excess are confined to the GDS which is a mathematical surface having zero thickness. The criticism[8] that the Gibbs approach is physically unrealistic arises from a complete misunderstanding of the nature of the Gibbs procedure. What the Gibbs method does is to recognise that the shape of the concentration profile which exists in the real system is not open to experimental determination, and provides a means of characterising the measurable global effect arising from non-uniformity near the surface. Secondly, some authors[8,9] advocate the use of a surface phase of volume V^{σ} defined by two arbitrary dividing surfaces placed in the homogeneous regions of the adjacent phases. This is said to be a 'more realistic' approach. It is then possible on this basis to define $\Gamma_2^{(1)}$ in the same form as equation(9). However, in describing other thermodynamic properties of the interface, a term in pV^{σ} appears; and it is then argued that this term is so small that it can be neglected. This means in effect that $V^{\sigma} \rightarrow 0$ and so the two dividing surfaces merge as one GDS. In the writer's view the surface phase model in this form (to be distinguished from the surface phase concept used later) has little to recommend it over the Gibbs method, which although it has been misinterpreted in the past and been the source of a good deal of confusion, is when correctly used a powerful and unambiguous tool.

Having established a method of defining the surface excess amount, it is clear that by an obvious extension of the procedure other surface excess quantities such as energy, entropy and free energy can be defined[10]:

$$U^\sigma \;=\; U - \overset{o}{u}{}'V' - \overset{o}{u}{}''V'' \tag{12}$$

$$S^\sigma \;=\; S - \overset{o}{s}{}'V' - \overset{o}{s}{}''V'' \tag{13}$$

$$F^\sigma \;=\; F - \overset{o}{f}{}'V' - \overset{o}{f}{}''V'' \tag{14}$$

where in each case the superscript ^{o} refers to the density of the property and denotes 'divided by volume': thus $\overset{o}{u}{}' = U'/V'$ etc. When taken relative to a GDS at the solid surface, and per unit area these surface excesses may be written $\overset{o}{u}{}^{\sigma(n)}$, $\overset{o}{s}{}^{\sigma(n)}$ and $\hat{f}^{\sigma(n)}$, where the use of a circumflex denotes 'divided by area'.

It is not our purpose here to develop the thermodynamics of interfaces in detail. The most important equation to follow from such a development is the Gibbs adsorption isotherm:[6,11]

$$-d\sigma \;=\; \sum_{i=2}^{c} \Gamma_i^{(1)} \, d\mu_i. \qquad (\text{const. } T), \tag{15}$$

where μ_i is the chemical potential of i, which is uniform throughout the bulk phase and the surface layer. In deriving this equation it is not necessary to introduce an operational definition of σ; it is sufficient to suppose that an increase in surface area dA at constant surface excess amounts and temperature leads to an increase in the energy of the system by σdA. For the lack of a better term σ is called the surface tension or interfacial tension of the solid/liquid interface. The importance of equation (15) is that it now provides us with a thermodynamically rigorous method of calculating changes in surface tension of the solid/liquid interface from a knowledge of $\Gamma_i^{(1)}$ and μ_i.

We have already seen that $\Gamma_2^{(1)}$ is linked to the experimental quantity $\Gamma_2^{(n)}$ so that, for a binary system,

$$-d\sigma \;=\; \frac{\Gamma_2^{(n)}}{x_1^\ell} \, d\mu_2. \tag{16}$$

Furthermore,

$$d\mu_2 = RT \, d \, \ln \, x_2^\ell \gamma_2^\ell , \qquad (17)$$

where γ_2^ℓ is the activity coefficient of component 2 in the bulk liquid. Inserting (17) into (16) and integrating, starting from pure component 2, we obtain[12]

$$\sigma - \sigma_2^* = -\frac{RT}{a_s} \int_{x_2^\ell}^{x_2^\ell = 1} \frac{n^\circ \Delta x_2^\ell / m}{x_1^\ell x_2^\ell \gamma_2^\ell} \, d \, (x_2^\ell \gamma_2^\ell) , \qquad (18)$$

where σ_2^* is the interfacial tension at the solid/pure liquid 2 interface, and σ is that at the solid/solution, of mole fraction x_2^ℓ, interface. If the integration is made across the whole concentration range, then $(\sigma_1^* - \sigma_2^*)$, the difference between the interfacial tensions of solid in contact with liquids 1 and 2 can be calculated.

If measurements are made at a series of temperatures then it is possible to calculate the following enthalpy and entropy quantities:[13,14]

$$\Delta_w \hat{h} - \Delta_w \hat{h}_2^* = \left[\frac{\partial}{\partial (1/T)} \left[\frac{\sigma - \sigma_2^*}{T} \right] \right]_{x_2^\ell} , \qquad (19)$$

$$\Delta_w \hat{s} - \Delta_w \hat{s}_2^* = -\frac{(\sigma - \sigma_2^*) - (\Delta_w h - \Delta_w \hat{h}_2^*)}{T} , \qquad (20)$$

where $\Delta_w \hat{h}$ and $\Delta_w \hat{s}$ are, respectively, the enthalpy and entropy of wetting (or immersion) per unit area of solid in the solution of mole fraction x_2^ℓ, and $\Delta_w \hat{h}_2^*$ and $\Delta_w \hat{s}_2^*$ are the corresponding quantities for the pure liquid 2.

<u>Experimental methods</u>

The definition of $\Gamma_2^{(n)}$ given above indicates the simplest and most direct method for its determination using a batch method in which solid and initial solution are mixed, allowed to come to equilibrium and the liquid analysed by an appropriate method. This procedure is tedious if an extensive study is to be made, and encounters many problems if accurate data over a range of temperature are required. The use of a closed system[15] (Figure 3), in which the solution is brought to equilibrium by circulation over the solid and compared with a reference system in a differential refractometer, overcomes most problems and can

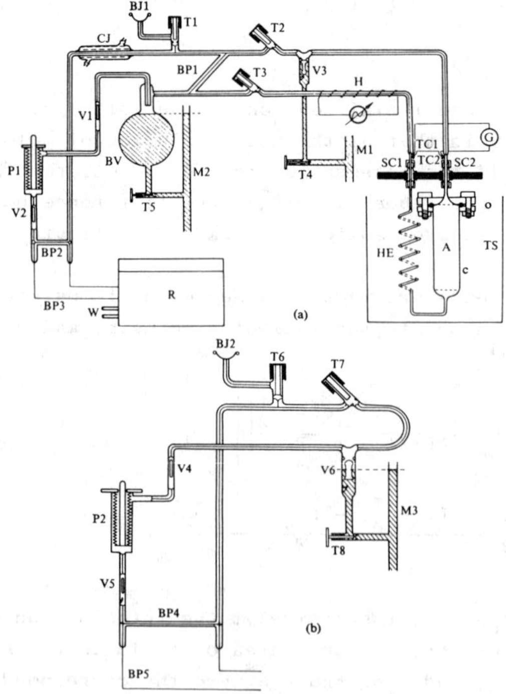

Figure 3: Differential adsorption apparatus; a, adsorption system; b, reference system. The adsorbent A is immersed in the thermostat TS and liquid is circulated by pumps P. The refractive index difference is recorded using the differential refractometer R. See ref.15 for further details (reproduced from ref.15 by permission).

be used to study the temperature coefficients of adsorption. A
third method uses a chromatographic type of technique in which a
solution of constant composition is passed through a column of
solid adsorbent[16]. Initially the concentration of the solution
leaving the bed is less than that of the feed, but as equilibrium
is approached the concentration in the effluent rises until it
equals that of the feed. The integrated amount of component 2
leaving the column is less than the total amount fed in by an
amount equal to the reduced surface excess of 2:

$$n_2^{\sigma(n)} = \int_0^{n^o} \Delta x_2^{\ell} \, dn, \tag{21}$$

where Δx_2^{ℓ} is here the difference between the mole fractions at the
inlet and outlet and n^o the amount of solution which has to be
passed to reduce Δx_2^{ℓ} to zero. Other chromatographic methods
of determining adsorption are possible[17], but most of them
involve the assumption that adsorption equilibrium is maintained
at each point in the column, and corrections have to be applied
for spreading of the chromatographic peak caused by diffusion and
other kinetic factors.

    Enthalpies of immersion determined directly by calorimetry
also play an important role in establishing the properties of
adsorbed systems. Most work has in the past been concerned with
immersion in pure liquids[18], but experiments on solutions, cover-
ing the full concentration range are equally important, especially
when adsorption data are available at only one temperature[19].
The experimental techniques available for measurements of this
kind have developed rapidly in the last decade, although it is
still unusual for a reproducibility of better than 1% to be
achieved. Immersion calorimetry can be supplemented by flow
displacement calorimetry and for some types of system this can
provide useful information[20].

Experimental Results

    Experimental measurements of adsorption from solution are best
represented in terms of <u>surface excess isotherms</u> (often called
<u>composite isotherms</u>) expressed as the specific reduced surface
excess ($n^o \Delta x_2^{\ell}/m$) as a function of x_2^{ℓ}. Isotherms of this kind can
adopt a variety of forms, the more important of which are the

so-called U and S-shaped isotherms (approximately types II and IV in the Schay-Nagy classification[21]), typical examples of which are shown in Figure 4. In the S-shaped isotherm the surface

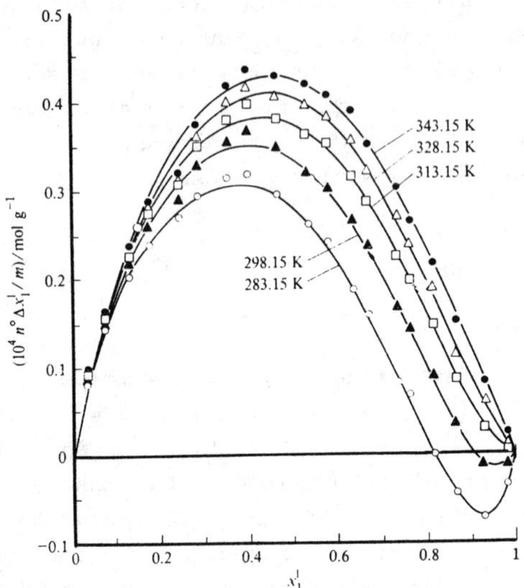

Figure 4: Surface excess isotherms for benzene (1) + n-heptane (2) + Graphon at indicated temperatures.

excess passes through zero at a particular concentration, and the preferentially adsorbed species changes from one component to the other. At zero surface excess the local composition of the system remains constant from the bulk up to the surface: this state is often called a <u>surface azeotrope</u>.

<u>Analysis of Results and Molecular Interpretations</u>

A full theory of adsorption from solution would involve the ability to calculate the form of the concentration profile up to the surface and to establish the orientation of the adsorbed molecules. Although progress towards this end is being made, it is still conventional to represent the concentration as a step function (Figure 5), dividing the system into a bulk phase, of mole fraction x_2^l, and a surface or adsorbed phase in which the mole fraction is x_2^σ. This model has many shortcomings, although it does have the merit of enabling the results of experiments to be

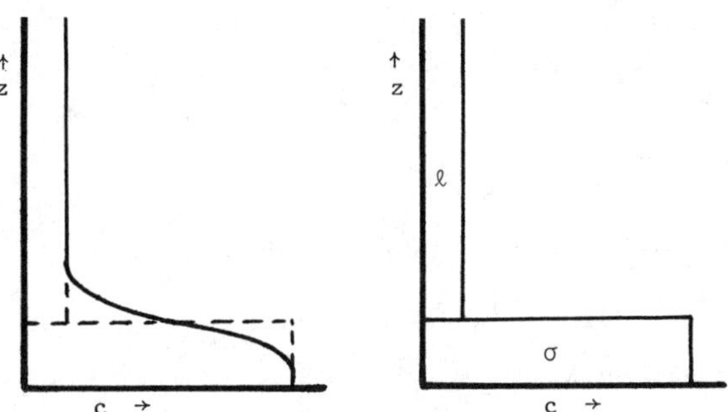

Figure 5: Concentration profile in real system and equivalent surface phase model.

systematised. In the simplest form of the theory the adsorbed layer is supposed to be one molecule thick, although in many real systems it is found that a thicker layer must be assumed if the analysis is to have any physical significance.

Adsorption equilibrium is set up when the desorption of one component, and the filling of the space left on the surface by the other component, is accompanied (at const. T) by zero change in free energy (Figure 6). This 'phase exchange' can be represented symbolically by the equation[22]

$$r(1)^{\sigma} + (2)^{\ell} \rightleftharpoons r(1)^{\ell} + (2)^{\sigma}, \tag{22}$$

where it is necessary to displace r molecules of component 1 to make room for one of component (2) on the surface. This

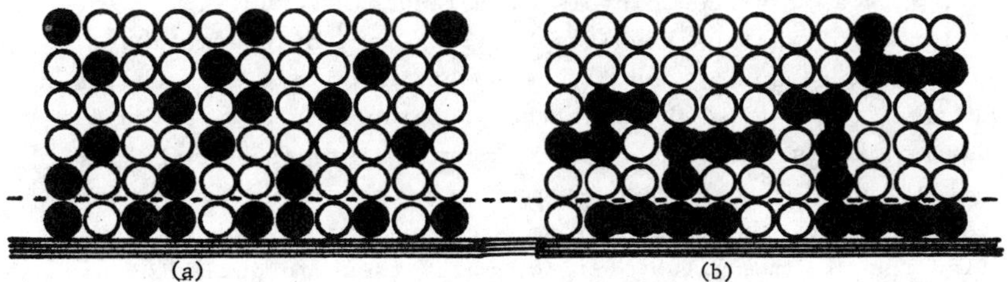

Figure 6: Adsorption as an exchange process (a) equal size molecules (b) different size molecules,"parallel layer" model.

equilibrium can be treated thermodynamically by considering the chemical potentials of adsorbed and bulk phase molecules, but it may also be looked upon in very simple terms using the law of mass action. On this basis it will be expected that in an ideal system equilibrium will be characterised by an adsorption equilibrium constant:

$$K = \left(\frac{x_1^{\ell}}{x_1^{\sigma}}\right)^r \frac{x_2^{\sigma}}{x_2^{\ell}} . \tag{23}$$

The thermodynamic approach links K_a with the standard free energy of the exchange process

$$\Delta G^{\ominus}_{ads} = -RT \ln K \tag{24}$$

where, it turns out,

$$\Delta G^{\ominus}_{ads} = (\sigma_2^* - \sigma_1^*)a_2, \tag{25}$$

where a_2 is the area occupied by component 2 on the surface. From this we see that if $\sigma_2^* < \sigma_1^*$ then component 2, that with the lower surface tension, is preferentially adsorbed ($K > 1$).

To apply equation (23) it is necessary to be able to calculate x_2^{σ}; this is given by[23]

$$x_2^{\sigma} = \frac{tx_2^{\ell} + a_1^{o}\,\Gamma_2^{(n)}}{t-(a_2^{o} - a_1^{o})\Gamma_2^{(n)}} \tag{26}$$

where t is the thickness of the adsorbed layer in molecular layers and a_1^{o}, a_2^{o} the molar areas of molecules 1,2 in a single layer; $a_2 = a_2^{o}/t$ since t layers of molecule 2 occupy a_2^{o}. Calculation of x_2^{σ} thus involves knowledge of, or assumptions regarding, t, a_1^{o} and a_2^{o} and of the specific surface area of the solid, a_s. One way of checking the acceptability of a given set of parameters is to recall that x_2^{σ} must satisfy the conditions $x_2^{\sigma} < 1, dx_2^{\sigma}/dx_2^{\ell} > 0$[24]. If the assumption of monolayer adsorption fails to satisfy these conditions, then values of t=2,3 are tested to find the minimum value of t for which these inequalities are satisfied[25]. For this purpose the areas a_1^{o}, a_2^{o} are usually estimated either from molecular models, or from the density of the bulk liquid regarded as a close-packed system.

If monolayer adsorption is assumed, and if the molecules are of essentially the same size (molar area = a) then equation 23 can be rearranged in the form[26]

$$\frac{x_1^{\ell}x_2^{\ell}}{n^{o}\Delta x_2^{\ell}/m} = \frac{m}{n^{\sigma}}\left\{x_2 + \left[\frac{1}{K-1}\right]\right\},\tag{27}$$

where $\dfrac{n^{\sigma}}{m}$ is the number of molecules which can be accommodated on the surface, namely

$$\frac{n^{\sigma}}{m} = \frac{a_s}{a}.\tag{28}$$

If this equation is followed (e.g. Figure 7) then the left

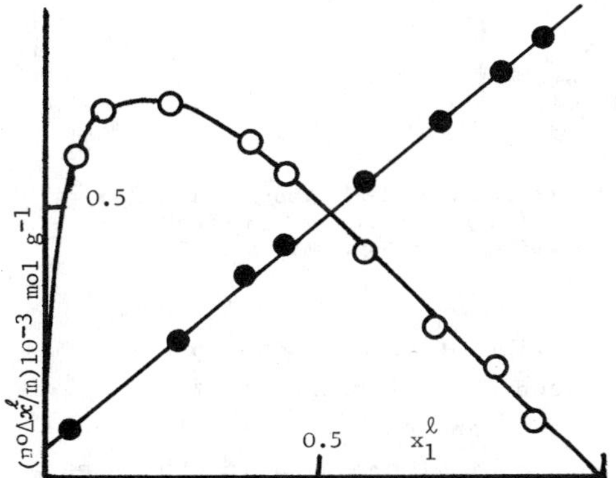

Figure 7: Adsorption by silica gel from benzene + cyclohexane mixtures.

O Surface excess isotherm
● Data plotted according to equation (27).

hand side will be a linear function of x_2^{ℓ} from which (n^{σ}/m) and hence a_2 and K can be determined. The form of the adsorption isotherm in this ideal case is shown in Figure 8 (a): two alternative graphical ways of finding n^{σ}/m when $K > 1$ are indicated, either by taking the limiting slope of the surface excess isotherm as $x_2^{\ell} \to 1$, or extrapolating the line joining the x_2^{ℓ} - axis at 0.5 and the maximum of the isotherm to the ordinate[26]. The use of adsorption from solution for the determination of surface area has been advocated[27], and in many instances the areas so obtained are in quite good agreement with those derived from BET-N$_2$ measurements. It must be remembered, however, that equation (27) applies only when molecules of equal size behave ideally in both bulk and surface regions. Despite this Schay and Nagy have been successful in using adsorption data even when an S-shaped isotherm is observed[28]. If such an isotherm has a

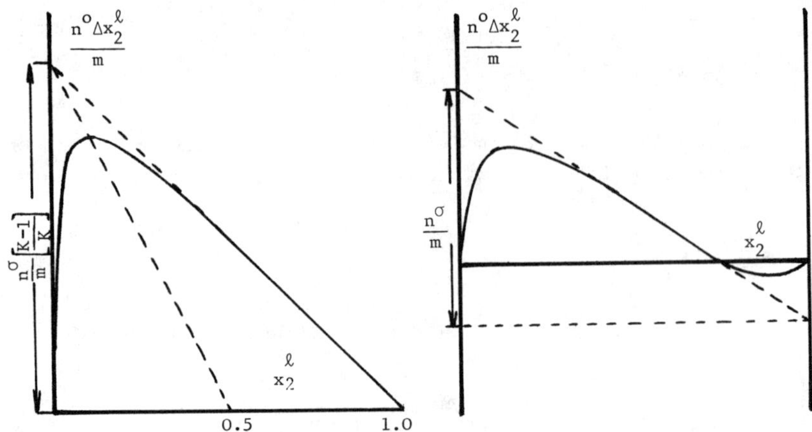

Figure 8: (a) Form of surface excess isotherm for ideal adsorption system
showing alternative graphical methods of finding monolayer capacity
(b) S-shaped surface excess isotherm showing Schay-Nagy method of finding
monolayer capacity.

linear section or extended inflection, then as shown in
Figure 8 (b) the difference between the intercepts on the
$x_2^\ell = 0$ and $x_1^\ell = 1$ ordinates, of the tangent at the inflection
point, is equal to n^σ/m. A check on the thermodynamic consistency
of the analysis of measurements on near-ideal systems can be
made by carrying out the integration of equation (8) across the
whole range to obtain $(\sigma_2^* - \sigma_1^*)$ and then using this to calculate
K by equation (24), and comparing this value with that derived
from equation (27).

In practice relatively few systems behave ideally. For non-
ideal systems one may adopt the formal device of introducing
activity coefficients, leading to the equilibrium constant in the
following form [29]:

$$K = \left\{\frac{x_1^\ell \gamma_1^\ell}{x_1^\sigma \gamma_1^\sigma}\right\}^r \left\{\frac{x_2^\sigma \gamma_2^\sigma}{x_2^\ell \gamma_2^\ell}\right\} . \tag{29}$$

There is no simple linearisation of this equation, but values of
the surface activity coefficients can be calculated from the
following thermodynamically derived equation:[30]

$$\ln \gamma_2^\sigma = \ln \left[x_2^\ell \gamma_2^\ell / x_2^\sigma\right] + a_2(\sigma_2^* - \sigma_1^*)/RT. \tag{30}$$

Once again calculations of this kind require values of x_2^σ and consequently depend on an assessment of the surface area, adsorbed layer thickness and values of a_1^o, a_2^o. While therefore it may sometimes be useful in simple cases to compare surface and bulk activity coefficients, such comparisons should, because of the number of parameters involved, be viewed with some caution.

One interesting example is, however, that in which a mixture of molecules of the same size exhibit regular solution behaviour in both bulk and adsorbed (monolayer) phases, characterised by an 'interaction parameter', α[29]. The resulting surface excess isotherms show, for certain combinations of $\ln K$ and α, azeotropic behaviour (Figure 9).

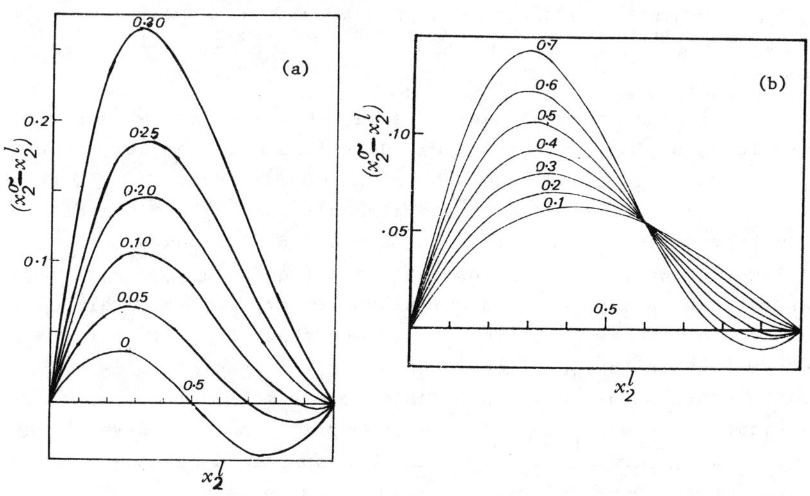

Figure 9: Adsorption from regular solutions $(x_2^\sigma - x_2^\ell)$ as functions of x_2^ℓ for (a) $\alpha/(2.303RT) = 0.50$ and a range of values of $\log K$ (on curves); (b) $\log K = 0.10$ and a range of values of $\alpha/(2.303RT)$ (on curves). Note: when t=1, and $a_1^o = a_2^o = a$, $(x_2^\sigma - x_2^\ell) = a\Gamma_2^{(n)}$.

Models for the adsorption of molecules of different size become rapidly more complex and it is less easy to compare them with experimental data. We may, for example, develop a theory in which the adsorbed chain molecules lie in a monolayer (parallel layer model)[31] (Figure 6b) or in which they can adopt various configurations in which at least one segment is in contact

with the surface[32] (Figure 10)   Earlier calculations were
limited to monomer + tetramer mixtures, but in recent developments
Fleer and Scheutjens [33] have been able to develop a theoretical
treatment which can be extended to molecules of any length.

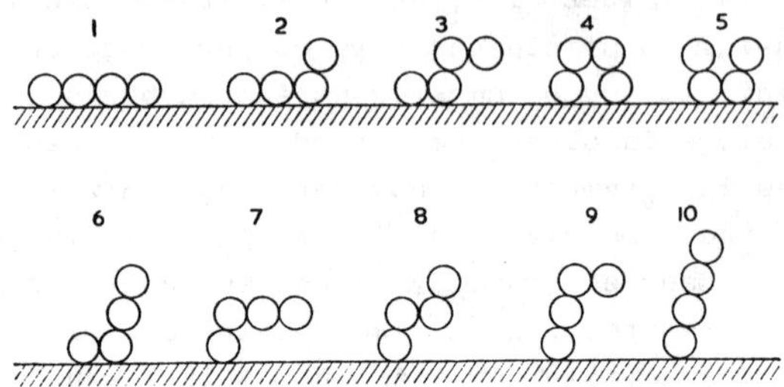

Figure 10: Possible configurations of flexible tetramer molecule with
at least one segment in contact with surface.

     Mention may also be made of experimental evidence, both from
enthalpies of immersion and from adsorption measurements, that
strong ordering of the adsorbed molecules can occur[34,35]. In the
case of long-chain n-alkanes and alcohols adsorbed by graphite
from heptane this structuring is manifested by a sharp inflection
in the isotherm. These phenomena still require a good deal more
experimental study before a definitive theory can be developed.
     So far it has been tacitly assumed that the solid surface is
homogeneous in the sense that the energy of adsorption is
independent of the location of an adsorbed molecule over the solid.
Experimental studies to validate adsorption theories have therefore usually employed solids such as graphitised carbon black
which, on the basis of other evidence, is believed to be
substantially homogeneous. But most surfaces of practical
importance are not homogeneous and it is important to assess the
influence which heterogeneity of the surface has on adsorption
phenomena. It turns out that this problem is not easily resolved,
for there is no unambiguous way of separating out the influence
of non-ideal behaviour in the surface region (arising from intermolecular forces in the surface region) from that of surface
heterogeneity. Thus it has sometimes been supposed[36] that the
existence of an adsorption azeotrope indicates the presence on the
surface of some sites which adsorb component 1 preferentially and

others which adsorb component 2, so that the surface excess
isotherm is characterised by two equilibrium constants K_1 and K_2
relating to the two types of site. However, it has already been
seen that imperfection in either or both the bulk and adsorbed
phases can lead to azeotropic behaviour, while often the chemical
nature of the surface is such that the existence of two such
different types of site seems very unlikely. Nevertheless, in
further development of work in this field surfaces of known,
and preferably controllable, heterogeneity should be investigated.

Enthalpy and Entropy Effects

An important insight into the factors which determine
adsorption equilibrium may be obtained by examining the
contributions from enthalpy and entropy effects. One might
expect that the preferential adsorption would be determined by
the relative values of the energies of interaction of the two
types of molecule with the solid surface, and that these energies
could be deduced from the enthalpies of immersion of the solid in
the two liquids. Thus one would expect the liquid, immersion of
the solid in which was the more exothermic, to be adsorbed
preferentially. While this is often the case it is by no means
universal as is revealed by the curves of $(\sigma - \sigma_2^*)$,
$(\Delta_w \hat{h} - \Delta_w \hat{h}_2^*)$ and $-T(\Delta_w \hat{s} - \Delta_w \hat{s}_2^*)$ as functions of $\overset{\ell}{x}_2$. The latter
two quantities are obtained from measurements of $(\sigma - \sigma_2^*)$ as a
function of temperature using equations (19) and (20) or from
direct calorimetric enthalpy of immersion data.

The sign of the adsorption of a given component is determined
by the slope $d(\sigma - \sigma_2^*)/dx_2$ of the surface tension curve: if this
slope is positive, the component 1 is preferentially adsorbed, if
negative, 2 is preferentially adsorbed. However, the surface
tension is the resultant of combining the enthalpy and entropy
curves as shown in some typical examples in Figure 11[37]. For the
(benzene + cyclohexane)/Graphon system (Figure 11 a) the surface
tension and enthalpy curves are virtually identical and the
entropy curve hardly deviates from zero over the whole concentration range ($\sigma - \sigma_2^*$ is independent of temperature within the
experimental accuracy). Here then the preferential adsorption is
determined solely by energetic factors. In the (cyclohexane +
n-heptane)/Graphon system (Figure 11 c) substantial entropy
effects work against the enthalpy, but are insufficient to change
the sign of the slope of the $(\sigma - \sigma_2^*)$ curve. In contrast, for

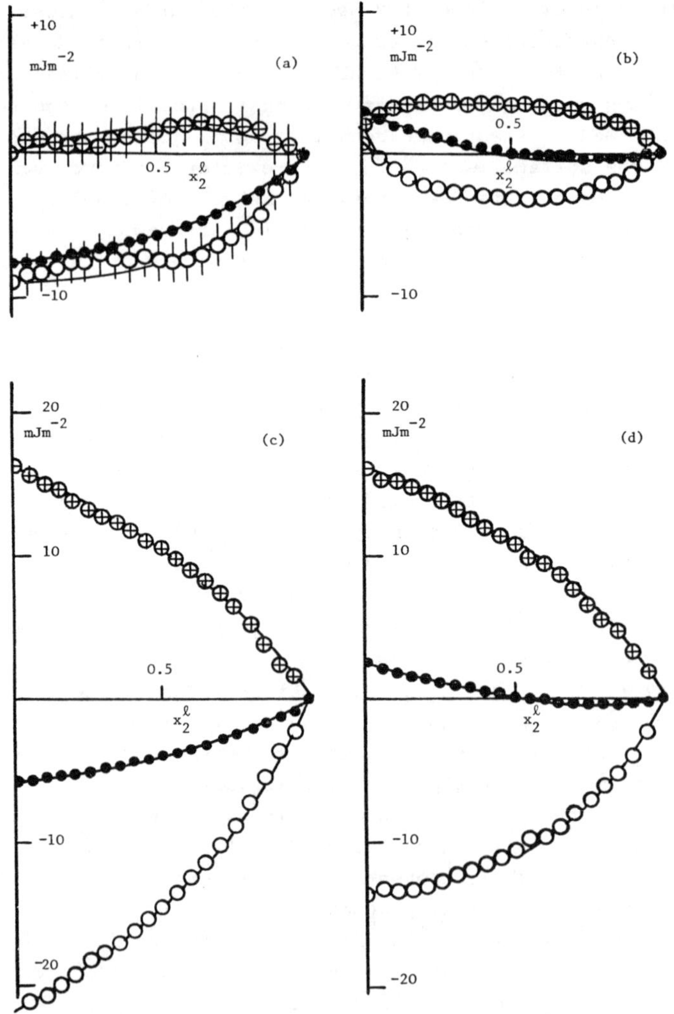

Figure 11: $(\sigma - \sigma_2^*)$ ● ; $(\Delta_w\hat{h} - \Delta_w\hat{h}_2^*)$ O ; and $-T(\Delta_w\hat{s} - \Delta_w\hat{s}_2^*)$ ⊕ as functions of x_2^ℓ for the systems (a) (benzene + cyclohexane)/Graphon, (b) (1,2 dichloroethane + benzene)/Graphon, (c) (n-heptane + cyclohexane)/Graphon (d) (n-heptane + benzene)/Graphon.

(benzene + n-heptane)/Graphon (Figure 11 d) the difference in the
enthalpies of adsorption is less than for cyclohexane + n-heptane
and is more than offset by entropy effects so that although
n-heptane is preferentially adsorbed at low mole fractions, at
higher concentrations benzene is preferentially adsorbed. Over
most of the concentration range the sign of the adsorption is
determined by entropic factors: qualitatively it appears that loss
of conformational entropy of n-heptane on adsorption opposes the
effect of the higher energy of adsorption of heptane compared
with that of benzene. It is of particular interest that in both
the last two systems the entropy curves are virtually identical
suggesting that the entropy changes on adsorption of heptane are
similar in the two instances. Finally, in the case of
(1,2-dichloroethane + benzene)/ Graphon[38] (Figure 11 b) not only
does $(\sigma - \sigma_2^*)$ change sign, but so also does $(\Delta_w \hat{h} - \Delta_w \hat{h}_2^*)$. Thus
even for a relatively simple system which does not depart
markedly from ideality there is evidence for complex
behaviour. These examples stress the importance of making a
detailed thermodynamic analysis before embarking on detailed
theories of adsorption from solution: in many instances the
theory is required to take account of complex structural changes
in the adsorbed layer whose existence would perhaps not have been
suspected in the absence of a thermodynamic study.

Some applications of adsorption from solution

Purification processes : adsorption from dilute solutions. The
practical applications of adsorption from solution for the
removal of impurities from liquids is widespread, and has become
particularly significant in the field of water treatment where
active charcoal is widely used. In this case, the concern is
mainly with dilute aqueous solutions of solids of limited
solubility. Under such circumstances the ideas developed earlier
take on a somewhat different aspect. First we notice that the
equilibrium constant K reduces to the form[39]

$$K' = \frac{x_2^{\sigma}}{x_2^{\ell}} \cdot \left(\frac{\gamma_2^{\sigma}}{\gamma_2^{\ell}}\right)_{\infty} , \tag{31}$$

since $x_1^{\ell}, x_1^{\sigma} \longrightarrow 1$ and the activity coefficients $\gamma_2^{\sigma}, \gamma_2^{\ell}$ tend to their
values at infinite dilution. The adsorption isotherm is thus

$$x_2^{\sigma} = K_h x_2^{\ell} \, , \tag{32}$$

i.e. a linear (Henry's law) isotherm is observed. K' is now related to the quantity $(\sigma_2^{\ominus} - \sigma_1^{*})$ where $\sigma_2^{\ominus}$ refers to a hypothetical standard state. If as often happens component 2 is incompletely miscible with 1, then the standard state of pure 2 cannot be realised experimentally. In the case of partially soluble solutes the method of representation of the experimental data is often modified. Thus by analogy with vapour adsorption, where isotherms are often plotted in terms of the relative pressure p/p^{*} (p^{*} = saturated v.p.), surface excess isotherms may be expressed as functions of $x_2^{\ell}/x_2^{\ell}(\text{sat})$, where $x_2^{\ell}(\text{sat})$ is the mole fraction of saturated solution[40]. Alternatively adsorption from dilute solution may be treated in terms of an extension of the Polanyi potential theory and the surface excess studied as a function of RT ln $x_2^{\ell}/x_2^{\ell}(\text{sat})$, the adsorption potential[41]. A further extension, adopted largely by the Russian school[42], is to extend the Dubinin–Radushkevich equation to solution adsorption. Here a linear relation is sought between the surface excess and RT $\left[\text{ln } x_2^{\ell}/x_2^{\ell}(\text{sat})\right]^{n}$ where n is often found to be 2. In some cases a correlation has been established between the parameters of such a linear relation and the properties of the adsorbed molecules[43]. This is an area to which further research should be directed since at the moment insufficient experimental data are available to enable a proper assessment of the suggested theoretical treatments to be made.

<u>Wetting and Displacement</u>

A completely different application is to the problems of wetting and liquid/liquid displacement. We return to equation (18) which enables $(\sigma_2^{*} - \sigma_1^{*})$ to be calculated from experimental adsorption data. First, we note that if substances 1,2 and 3 are mutually miscible in all proportions, then experiments on the three binary mixtures should satisfy the identity

$$(\sigma_1^{*} - \sigma_2^{*}) + (\sigma_2^{*} - \sigma_3^{*}) + (\sigma_3^{*} - \sigma_1^{*}) = 0 \, . \tag{33}$$

We have in this equation an important test both of the reliability of experimental data and of the computational methods used in their analysis[44].

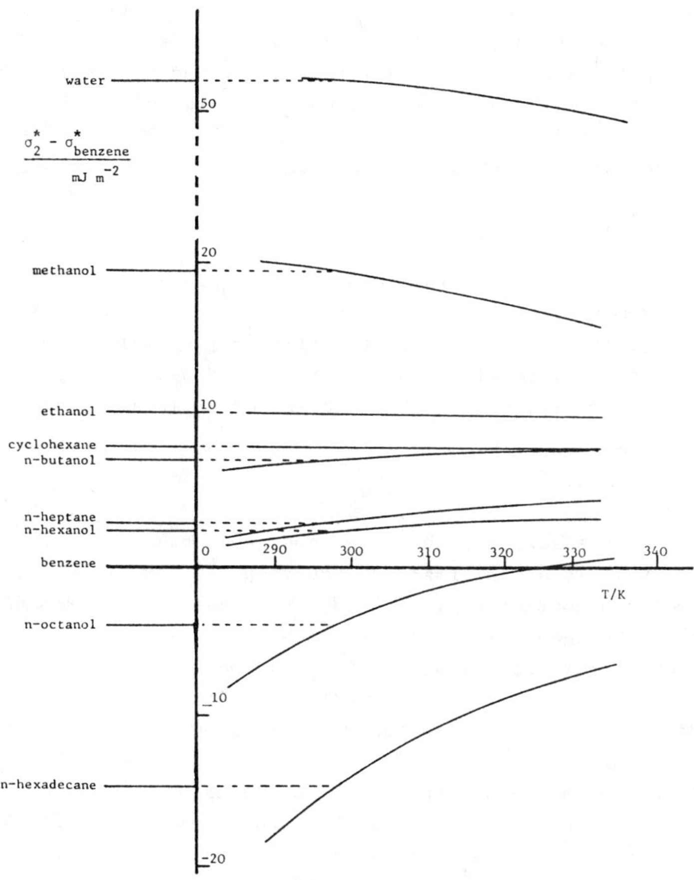

Figure 12: Surface tensions of liquid 2/graphite interface relative to that of the benzene/graphite interface

Left hand: at 298 K   Right hand: as function of temperature

The data upon which this figure is based come from references 15,25, R. Bown, Ph.D. thesis Bristol 1973; R.W. Smith, Ph.D. thesis Bristol,1978; C.E. Brown, unpublished; A.J.P. Fletcher, unpublished.

However, if only two of these pairs are completely miscible whereas the third (say 1 + 2) pair is completely immiscible, then measurements on (1 + 3) and (2 + 3) provide a means of evaluating $(\sigma_2^* - \sigma_1^*)$. We recall that if σ_2^* and σ_1^* are the surface tensions of pure 2 and pure 1 in contact with the solid surface, then they are related to the contact angle (θ) at the 1/2/solid line of contact, measured through liquid 1, by Young's equation.

$$\sigma_2^* - \sigma_1^* = \sigma_{12} \cos\theta, \tag{34}$$

where σ_{12} is the interfacial tension of liquid 1/liquid 2. Thus adsorption measurements provide an indirect method of evaluating $\cos\theta$ [45]. This procedure is potentially very useful where the solid is in the form of a powder, compressed pellet or porous body. However, Young's equation has an operational meaning only when

$$\frac{\sigma_2^* - \sigma_1^*}{\sigma_{12}} < 1.$$

It is sometimes supposed that when $\cos\theta = 1$, $\sigma_2^* - \sigma_1^* = \sigma_{12}$. However, this is an unwarranted deduction. An experimentally observed zero contact angle includes the case in which component 2 penetrates between component 1 and the solid; i.e. component 2 displaces 1 from the surface of 2. This occurs when $[(\sigma_2^* - \sigma_1^*) - \sigma_{12}]$ is positive. This quantity may be called the <u>spreading pressure</u>, or with opposite sign the <u>spreading tension</u>, σ_{spr}.

We now have the possibility of building up a scale of liquid/solid surface tensions which can be used to estimate contact angles or, when $[(\sigma_2^* - \sigma_1^*) - \sigma_{12}]$ is positive, the displacement characteristics of the 1,2 couple. This is illustrated for the graphite interface in Figure 12 where benzene is chosen as a reference liquid. The relative wettability of graphite by water and an organic liquid with which it is immiscible can be assessed by reading off the difference $(\sigma_{water}^* - \sigma_{organic}^*)$ and comparing this value with the interfacial tension σ_{wo} at the water/organic inter-face [46]. If $(\sigma_w^* - \sigma_o^*) < \sigma_{wo}$ the w/o interface will make a finite contact angle with graphite, whereas when $(\sigma_w^* - \sigma_o^*) > \sigma_{wo}$ the organic liquid will displace water from a graphite surface. Thus for heptane/water/graphite the former situation arises with a predicted θ of about $7°$: experimental observation on a freshly cleaned graphite surface gives values of $6\text{-}9°$ [47]. On the other hand for the benzene/water/graphite system it is predicted that benzene will displace water: experimentally $\theta = 0^{47}$. It may be noticed that as the length of an alkane chain increases, the tendency to displace water also increases; this also is confirmed by observation. Clearly information of this kind, obtained for other solids, is of considerable potential value in discussing wettability and displacement phenomena in, for example, porous media.

General References

A D.H. Everett in Specialist Periodical Reports, Colloid Science, D.H. Everett(ed.)Vol 1,Chapter 2,Chemical Society London (1973) Note errata published in Ref.B,p.314.

B C.E. Brown and D.H. Everett, Specialist Periodical Reports, Colloid Science, D.H. Everett(ed.)Vol 2, Chapter 2, Chemical Society London (1975).

C D.H. Everett and R.T. Podoll, Specialist Periodical Reports, Colloid Science, D.H. Everett(ed.) Vol 3,Chapter 2, Chemical Society London (1979).

D R. Defay, I. Prigogine, A. Bellemans and D.H. Everett, Surface Tension and Adsorption,Longmans 1965.

Specific References

1. J. Lyklema,this volume Chapter 3.

2. D.H. Napper,this volume, Chapter 5.

3. S.G. Ash, D.H. Everett and C.J. Radke, J.C.S. Faraday II, 1973, 69,1256;
   D.H. Everett and C.J. Radke, A.C.S. Symposium Series No.8, K.L. Mittal(ed.),1975,1;
   D.H. Everett, Pure Appl.Chem. ,1976,48,419;
   D.G. Hall,J.C.S. Faraday II,1972,68,2169;

4. International Union of Pure and Applied Chemistry,Div. of Physical Chemistry, Manual of Symbols and Terminology, Appendix II,Part I; Pure Appl.Chem.,1972,31,579.

5. Ref.A,p.50.

6. J.W. Gibbs, Collected Works,Longmans Green, New York,London, Toronto,1928,p.219.

7. Ref D,p.26.

8. e.g. E.A. Guggenheim, Trans.Faraday Soc. 1940,36,397;
    Thermodynamics, 5th Ed. North Holland, Amsterdam 1967,46,207

9. J.E. Verschaffeldt,Bull.Ac.Roy.Belg.(Cl Sc.)1936,22,373,390,402.

10. Ref.D,p.23; Ref.6, p.224.

11. Ref.D,p.85.

12. Ref.A,p.58 (note signs on r.h.s. of equ.(38),39) should be
    reversed).

13. G. Schay,J. Colloid Interface Sci,1973,42,478; S. Sircar,
    J. Novosad and A.L. Myers, I and EC Fundamentals, 1972,11, 249.

14. Ref.B, p.54; Ref.C, p.65.

15. S.G. Ash, R. Bown and D.H. Everett,J.Chem.Thermodynamics, 1973
    5,239. cf.E.Kurbanbekov, O.G. Larionov, K.V. Chmutov and
    M.D. Yudelevich,Zhur.Fiz.Khim. 1969,43,1630 (Russ.J.Phys.Chem.,
    1969,43,916).

16. S.C. Sharma and T. Frost, J. Colloid Interface Sci.,1973,43,36.

17. e.g. E. Glückauf, Nature, 1945,156,748; S.I. Zverev, O.G.
    Larionov and K.V. Chmutov, Zhur.Fiz.Khim.1974,48,1556,2101,2126.
    (Russ.J.Phys.Chem.,1974,48,916,1244,1261); H.L. Wang,J.L. Duda
    and C.J. Radke,J.Colloid Interface Sci.,1978,66,153.

18. A.C. Zettlemoyer and K.S. Narayan in 'The Solid-Liquid
    Interface', E.A. Flood(Ed.),Dekker,New York,1967,Vol 1,Chap.6;
    D.H. Everett, G.H. Findenegg and P.J. Gram,J.Chem.Thermo-
    dynamics,1969,1,573; J.H. Clint,J.S.Clunie, J.F. Goodman and
    J.R. Tate, Nature,1969,223,51.

19. G.D. Parfitt and M.W. Tideswell, J.Colloid Interface Sci.,
    1981,79,518.

20. e.g. A.J. Groszek,Proc.Roy.Soc.,1970,A314,473.

21. L.G. Nagy and G. Schay, J.Chim.Phys.1961,140.

22. D.H. Everett, Trans.Faraday Soc.,1964,60,1803; ibid.,1965,61,
    2478. cf. Ref.A,p.53.

23. Ref.A, p.67.

24. A.I. Rusanov, 'Phase Equilibrium and Surface Phenomena',
    Chimica,Leningrad 1967,Chap· VI; cf. ref A,p.66.

25. C. Brown,D.H. Everett and C.J. Morgan, J.C.S. Faraday I,1975,
    71,883.

26. Ref.A,p.55.

27. G. Schay in Surface Area Determination,D.H. Everett and
    R.H. Ottewill eds. p.273, Butterworths,London 1970.

28. G. Schay and L.G. Nagy, Acta Chim.Hung.1966,60,1803.

29. D.H. Everett, <u>Trans.Faraday Soc.</u>, 1965,<u>61</u>,2478.

30. G. Schay, in Surface and Colloid Science, E. Matijevic,Ed., Interscience, New York, Vol.2, 1969,p.155.see equation (120).

31. S.G. Ash, D.H. Everett and G.H. Findenegg,<u>Trans.Faraday Soc.</u>, 1968,<u>64</u>,2639.

32. S.G. Ash, D.H. Everett and G.H. Fir.denegg,<u>Trans.Faraday Soc.</u> 1968,<u>64</u>,2645;<u>ibid</u>, 1970,<u>66</u>,708.

33. J.M.H.M. Scheutjens and G.J. Fleer, <u>J.Phys.Chem.</u>,1979,<u>83</u>,1619; <u>ibid.</u>,1980,<u>84</u>,178.

34. D.H. Everett,<u>Israel J Chem.</u>,1975,<u>14</u>,267.

35. D.H. Everett,<u>Progr.Colloid and Polymer Sci.</u>,1978,<u>65</u>,103.

36. M. Siskova and E. Erdös, <u>Coll.Czech.Chem.Comm.</u>,1960,<u>25</u>,2599.

37. D.H. Everett, <u>J.Phys.Chem.</u>,1981,<u>85</u>, 3263.

38. D.H. Everett and R.T. Podoll, <u>J. Coll.Interface Sci.</u>,1981,<u>82</u>, 14.

39. V.A. Avramenko, V.Yu.Glushchenko and V.P. Shesterkin,<u>Zhur. Fiz.Khim.</u>,1977,<u>51</u>,2304 (<u>Russ.J.Phys.Chem.</u>1977,<u>51</u>,1348).

40. e.g. R.S. Hansen and R.P. Craig, <u>J.Phys.Chem.</u>, 1954, <u>58</u>,211.

41. M. Manes and L.J.E. Hofer, <u>J.Phys.Chem.</u>,1969,<u>73</u>,584.

42. A.M. Koganowiski and T.M. Levchenko,<u>Zhur.Fiz.Khim.</u>,1972,<u>46</u>, 1789 (<u>Russ.J.Phys.Chem.</u>,1972,<u>46</u>,1025);A.M. Stadnik and Yu.A. Eltekov, <u>Zhur.Prikl.Khim.</u> 1975,<u>48</u>,186.

43. e.g. A.M. Stadnik and Yu.A. Eltekov, <u>Zhur.Fiz.Khim</u>,1978,<u>52</u>, 1795(<u>Russ.J.Phys.Chem.</u>,1978,<u>52</u>,1041).

44. Ref.A,p.62, Ref. C, pp.86,127,135.

45. D.H. Everett, <u>Pure Appl. Chem.</u>,1980,<u>52</u>,1279.

46. D.H. Everett, <u>Pure Appl. Chem.</u>, 1981,<u>53</u>,2181.

47. A.J.B. Fletcher, Ph.D.Thesis, Bristol, 1982.

5
Polymeric Stabilization

By D. H. Napper

DEPARTMENT OF PHYSICAL CHEMISTRY, THE UNIVERSITY OF SYDNEY, SYDNEY, NEW SOUTH WALES, 2006, AUSTRALIA

Introduction

The stabilization of colloidal dispersions by naturally occurring polymers has been exploited continuously for some five millennia since the production in ancient Egypt of 'instant' ink for writing on papyrus.[1] This ink was prepared by mixing lamp black, obtained by combustion, with a solution of a natural steric stabilizer (e.g., casein from milk, egg albumin or gum arabic). The carbon black was then moulded into a pencil shape and dried. When the ink was required, the mould was simply dipped into water and the sterically stabilized carbon black particles redispersed spontaneously.

Nowadays stabilization by either natural or synthetic polymers is exploited in a wide range of industrial products: paints, glues, inks, pharmaceutical and food emulsions, detergents, lubricants, etc. Moreover, steric stabilization is evident in many biological systems, such as milk and blood, and important in a broad range of industrial and other processes (e.g., water treatment, soil stability, coal washing, oil recovery, etc.).

Steric stabilization possesses several advantages over other methods of imparting stability. These include its relative insensitivity to high concentrations of electrolytes (*c.f.* electrostatic stabilization); the fact that high solids dispersions display relatively low viscosities; and the equal effectiveness of steric stabilization in both aqueous and nonaqueous dispersion media. Good freeze-thaw stability is another feature of sterically stabilized dispersions.

The Need for Polymer Stabilization

Naked uncharged colloidal particles at the concentrations normally of interest undergo rapid coagulation, the number of separate particles being halved in a matter of only a few seconds. This instability is a consequence of the 'stickiness' of Brownian collisions between colloidal particles. This stickiness originates in the attractive van der Waals forces operative between such particles. These attractive interactions are of a relatively long-range character (typically of order 10 nm) and arise from the London dispersion forces.[2] The latter have a quantum mechanical origin and are the same forces that are responsible for the liquefaction of the rare gases at low temperatures. A simple classical picture of the origin of this attraction would implicate the correlations in the positions of the electrons relative to their respective nuclei. These generate attractive instantaneous dipoles which, of course, time average to zero (permanent) dipole moment.

Types of Polymer Stabilization

To generate stable dispersions, it is necessary to provide a repulsive interaction that outweighs the van der Waals attraction between the colloidal particles. This can at present be achieved in a practically useful fashion by only a remarkably small number of different mechanisms:

(i) electrostatic stabilization, which exploits the coulombic repulsion operative between charged, colloidal particles and their respective double layers;

(ii) polymeric stabilization, which for nonionic polymers can be accomplished in at least two distinct ways:

A. Steric stabilization, whereby stability is imparted by polymers adsorbed or attached to the colloidal particles;

B. depletion stabilization, which is imparted by polymer chains in free solution.

It is, of course, possible to use a combination of the foregoing methods (e.g., electrostatic plus steric, also termed electro-

steric, stabilization; steric plus depletion stabilization, etc.)
Note, too, that other methods of stabilizing colloidal particles
have been proposed (e.g., solvation effects) but these phenomena,
if they exist, cannot as yet be exploited practically.

Steric Stabilization[3-7]

The Best Stabilizers. It has long been recognized by manufac-
turers of nonionic polymeric stabilizers that the best steric
stabilizers are amphipathic in character (the hydrophilic -
lipophilic balance or HLB rating). These take the form of block
or graft copolymers composed of nominally insoluble anchor
polymers attached to nominally soluble stabilizing moieties.[8]
Simple examples of effective aqueous steric stabilizers are
poly(styrene -*b*- oxyethylene) or poly(vinyl alcohol-*g*-vinyl
acetate) whereas typical stabilizers for many nonaqueous dispersions are poly(acrylonitrile-*b*-styrene) and poly(vinyl chloride-
g-isobutylene).

The role of the anchor polymer is to prevent desorption of
the stabilizer under the stress induced by the close approach
of a second particle in a Brownian collision. Since such stress
could also induce lateral movement of the stabilizing chains,
complete surface coverage of the colloidal particles is also
necessary to optimize stability. This additionally minimizes
the possibility of the stabilizing moietiesattached to one
particle adsorbing on a second particle, and thus inducing
bridging flocculation.

The foregoing is not meant to imply that homopolymers cannot
impart steric stabilization.[9] Homopolymers are relatively
ineffectual, however, because of the conflicting requirements
that the dispersion medium be a poor solvent to ensure strong
adsorption of the stabilizer onto the particle but a good solvent
to impart effective steric stabilization. As a result, the
instability of dispersions stabilized by homopolymers often
arises from the lateral movement and/or desorption of the
stabilizing chains.

Phenomenology of Flocculation. Thermodynamic factors limit the
stability of sterically stabilized dispersions and emulsions
provided that the anchoring is adequate and the surface coverage

complete. This limit has been shown experimentally to be
independent of the nature of the anchor polymer (if effective)
and, commonly, the nature and size of the colloidal particles.[10-13]
Exceptions to this general pattern of behaviour may occur with
large particles coated by very thin steric layers. Nevertheless,
the thermodynamic limit to the stability of many sterically
stabilized systems is determined primarily by the chemical nature
of the stabilizing moities and the dispersion medium.

   Instability is induced by decreasing the solvency of the
dispersion medium for the stabilizing moieties. This may be
achieved in at least three different ways: by altering the
temperature;[10-19] by changing the pressure;[20] and by adding a
(miscible) non-solvent for the stabilizing moieties to the dispersion medium.[11] The transformation from long-term stability to
catastrophic flocculation occurs abruptly for sterically stabilized
dispersions at the critical flocculation point (cfpt) (see
Fig. 1)[10] The critical flocculation point may be a critical

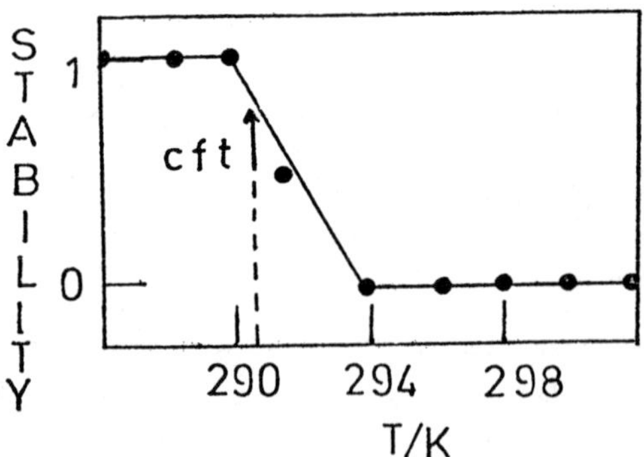

Fig. 1 Transformation from long-term stability to catastrophic
flocculation for poly(oxyethylene) stabilized latices in
0.39 M $MgSO_4$.

flocculation temperature (cft), pressure (cfp) or volume (cfv)
(for a nonsolvent that induces flocculation). Flocculation is
usually readily reversible. If flocculation is induced by
heating, the cfpt is referred to as an 'upper critical floccula-

tion point' (ucft) whereas if flocculation is induced by cooling, it is termed a 'lower critical flocculation point' (lcft).

Identification of the Critical Flocculation Point. Irrespective of whether flocculation is induced by a change in temperature or pressure, or by the addition of non-solvent, many sterically stabilized systems display a strong correlation between the critical flocculation point and the theta (or Flory)-point for the stabilizing moieties free in a solution of the dispersion medium. Some results supporting this correlation between the cfpt and the θ-temperature are presented in Table 1. The strong correlation was observed for both flocculation by heating (ucft) and cooling (lcft). These results were obtained by three different groups: Croucher and Hair in Canada, Dawkins and Taylor in England and Napper and co-workers in Australia. Other results that demonstrate a strong correlation between the critical flocculation pressure and the θ-pressure,[20] and the critical flocculation volume and θ-volume[11] have also been reported.

One test for the internal consistency of this correlation between the cft and θ is the molecular weight dependence of the cft. The θ-point is a property of the segment-solvent interactions and, as such, is independent of the molecular weight for high polymers. Accordingly, the cfpt should be independent of the molecular weight of the stabilizing moieties. The data presented in Table 1 adequately confirm this prediction.

The most elegant studies reported on the molecular weight dependence of the cft are those of Dawkins and Taylor.[18] These authors studied monodisperse poly(dimethylsiloxane)(PDMS) of different molecular weight (see Table 1), ranging from 3 200 to 48 000, as the stabilizing moieties. The cfts of the resulting dispersions were identical within experimental error (Table 1). These results, together with the other data displayed in Table 1, disprove the assertion by Vincent[23] that the correlation between θ and the cft is only valid in the limit of high molecular weight stabilizing moieties. The absence of any significant limit for high polymers is scarcely surprising if it is recalled that attachment of the stabilizing chains to the latex particles effectively transforms them into chains of infinite molecular weight by removal of the ideal (and strongly molecular weight

Table 1

Comparison of Theta Temperatures with Critical Flocculation Temperatures

Stabilizer	Molecular Weight	Dispersion Medium	cfpt/K	u/l	θ/K	Ref.
Poly(ethylene oxide)	10 000	0.39 M $MgSO_4$	318±2	ucft	315±3	10
Poly(ethylene oxide)	96 000	0.39 M $MgSO_4$	316±2	ucft	315±3	10
Poly(ethylene oxide)	1 000 000	0.39 M $MgSO_4$	317±2	ucft	315±3	10
Poly(acrylic acid)	9 800	0.2 M HCl	287±2	lcft	287±5	19
Poly(acrylic acid)	51 900	0.2 M HCl	283±2	lcft	287±5	19
Poly(acrylic acid)	89 700	0.2 M HCl	281±1	lcft	287±5	19
Poly(vinyl alcohol)	26 000	2 M NaCl	302±3	ucft	300±3	22
Poly(vinyl alcohol)	57 000	2 M NaCl	301±3	ucft	300±3	22
Poly(vinyl alcohol)	270 000	2 M NaCl	312±3	ucft	300±3	22
Polyacrylamide	18 000	2.1 M $(NH_4)_2SO_4$	292±3	lcft	–	19
Polyacrylamide	60 000	2.1 M $(NH_4)_2SO_4$	295±5	lcft	–	19
Polyacrylamide	180 000	2.1 M $(NH_4)_2SO_4$	280±7	lcft	–	19
Polyisobutylene	23 000	2-methylbutane	325±1	ucft	325±2	21
Polyisobutylene	150 000	2-methylbutane	325±1	ucft	325±2	21
Polyisobutylene	760 000	2-methylbutane	327	ucft	318	13
Polyisobutylene	760 000	2-methylpentane	381	ucft	376	13
Polyisobutylene	760 000	2-methylhexane	423	ucft	426	13
Polyisobutylene	760 000	3-ethylpentane	463	ucft	458	13
Polyisobutylene	760 000	cyclopentane	455	ucft	461	13
Poly(α-methyl styrene)	9 400	n-butyl chloride	254±1	lcft	263	14
Poly(α-methyl styrene)	9 400	n-butyl chloride	403±1	lcft	412	14
Polystyrene	110 000	cyclopentane	410	ucft	427	17
Polystyrene	110 000	cyclopentane	280	ucft	293	17
PDMS	3 200	n-heptane/ethanol	340	lcft	340±2	18
PDMS	11 200	(51:49 v/v)	340	lcft	340±2	18
PDMS	23 000		341	lcft	340±2	18
PDMS	48 000		338	lcft	340±2	18

dependent) contribution to their chemical potential in solution.

The data shown in Table 1 also contradict the postulate of Cairns and Neustadter[24, 25] that flocculation 'occurs under conditions which lead to phase separation of the stabilizing polymer in free solution at the same concentration as that of the adsorbed polymer'. This concept must be invalid, if only because of the indeterminancy of the concentration of the adsorbed polymer. The segmental concentration is usually highly variable within any given steric barrier, even if monodisperse in molecular weight. The data of Dawkins and Taylor on the cfts of latices stabilized by different molecular weight PDMS effectively disprove any hypothesis linking phase separation in free solution (which is strongly molecular weight dependent at these molecular weights) to the cfpt. Again, attachment of the chains to the particles removes the ideal contribution to their chemical potential and not surprisingly destroys any such nexus.

<u>Particle Concentration Dependence</u>. Everett and Stageman have claimed that the measured cft is strongly dependent upon the particle number concentration.[26, 27] This is not, however, borne out by the results of all workers: both Croucher and Hair[13] and Dawkins and Taylor[18] have reported that the cfts of their nonaqueous dispersions were insensitive to the particle number concentration. Moreover, Fig. 2 displays some results obtained by Feigin *et al.*[28] for latices stabilized by poly(oxyethylene) in 0.39 M $MgSO_4$. Within the limits of experimental error, the cft was found to be virtually independent of the particle number concentration over a concentration range of several orders of magnitude.

Feigin *et al.*[28] have shown, however, that a small concentration dependence of the cft could well arise from the loss of configurational entropy by the particles on flocculation. This loss of configurational entropy, which would correspond to a free energy change of order 10 kT at most for the usual particle concentrations of interest, is dependent upon the particle concentration. The effect is unlikely to be appreciable in aqueous systems because the relatively small partial molar volume of water renders unimportant such a small change in the interactional free energy. Any concentration dependence of the cft is more likely to be detected in nonaqueous systems where the

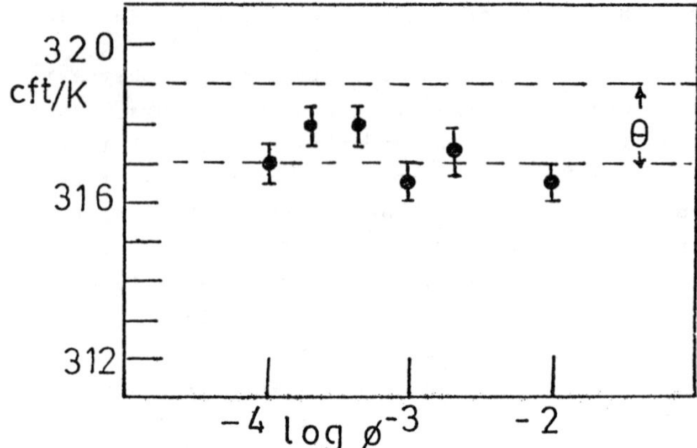

Fig. 2 The dependence of the cft upon the particle number concentration for a poly(oxyethylene) stabilized latex in 0.39 M $MgSO_4$.

molar volume of the dispersion medium is larger. Whether the
particle concentration dependence of the cft reported by Everett
and Stageman[26, 27] arose from the configurational entropy of their
latex particles is uncertain because the procedures they employed
in preparing their dispersions may well have led to the appearance
of artifacts arising from the contamination of the dispersion
medium.

Note that it is possible to show theoretically that the
transition from long-term stability to effectively complete
flocculation occurs, at least for aqueous dispersions, over a
very narrow temperature range (several Kelvin at most).[28] This
ensures that the effects of any particle concentration dependence
will be muted at the cft.

Classification of Sterically Stabilized Dispersions. The temperature dependence of the stability of sterically stabilized dispersions provides clues to the thermodynamic factors that control
stability *near to the cft*.[10] In principle, most, if not all,
sterically stabilized dispersions can be flocculated both by
heating and cooling. In practice, however, both of these cfts
are not always readily accessible.

The temperature dependence can be understood in terms of

$$\Delta G_T = \Delta H_T - T\Delta S_T \qquad (1)$$

where ΔG_T = free energy of close approach of sterically stabilized particles and ΔH_T and ΔS_T are the corresponding enthalpy and entropy changes so that $\partial(\Delta G_T)/\partial T = -\Delta S_T$. Possible combinations of signs for steric stabilization are shown in Table 2 and presented schematically in Fig. 3.

Table 2

Possible Combinations of Signs of the
Thermodynamic Parameters

ΔH_T	ΔS_T	$\|\Delta H\|/\|T\Delta S_T\|$	Type	Flocculation
+	+	≥ 1	enthalpic	heating
−	−	≤ 1	entropic	cooling
+	−	$\gtrless 1$	combined	inaccessible

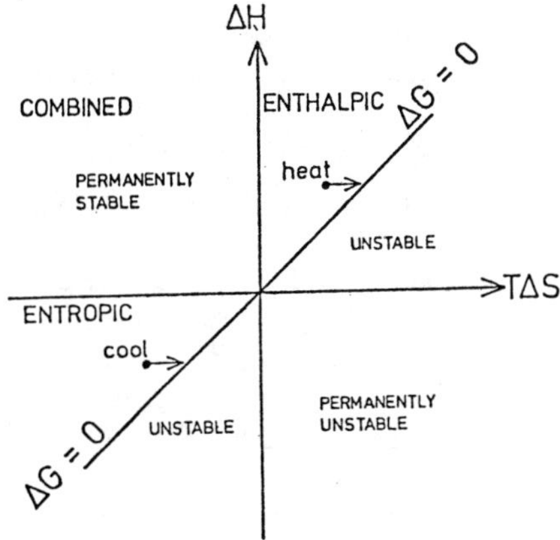

Fig. 3   Schematic representation of the thermodynamic factors controlling steric stabilization.

Note that at ambient temperature and pressure, entropic stabilization is more common in nonaqueous dispersion media whereas enthalpic stabilization is probably more common in aqueous dispersions. However, both entropic and enthalpic stabilization occur in both aqueous and nonaqueous media. Indeed, most, if not all, dispersions should in principle display both entropic and enthalpic stabilization under different conditions of temperature and pressure. Moreover, as shown in Fig. 3, it must pass through a domain of combined enthalpic plus entropic stabilization in converting from one stability type to the other.

Table 3 presents the classification of some representative aqueous and nonaqueous dispersions at room temperature and pressure (and also near to their cft).

Table 3

Classification of Some Sterically Stabilized
Dispersions at Room Temperature and Pressure

Stabilizer	Type	Dispersion Medium	
		Example	Stabilization
poly(oxyethylene)	aqueous	0.39 M $MgSO_4$	enthalpic
poly(vinyl alcohol)	aqueous	2 M NaCl	enthalpic
poly(methacrylic acid)	aqueous	0.02 M HCl	enthalpic
poly(acrylic acid)	aqueous	0.2 M HCl	entropic
poly(acrylamide)	aqueous	2.1 $(NH_4)_2SO_4$	entropic
polystyrene	nonqueous	cyclopentane	entropic
polyisobutylene	nonqueous	2-methylbutane	enthalpic

<u>The Unimportance of van der Waals Forces in Incipient Flocculation.</u>
For many sterically stabilized dispersions, van der Waals forces are too small to be responsible for flocculation near to the θ-point. In this respect, steric stabilization is markedly different from electrostatic stabilization. Only with large particles and thin steric barriers do the van der Waals forces manifest themselves in flocculation. Flory was the first to appreciate that the 'stabilizing' moieties may play a dual role in sterically stabilized dispersions in the sense that they

generate repulsive interactions, and thus stability, in dispersion media that are better solvents than θ-solvents whilst being attractive ('sticky'), and thus promoting flocculation, in worse than θ-solvents.

Qualitative Discussion of the Origins of Steric Stabilization

Croucher and Hair[13-17] have pointed out that a manifold of phenomena give rise to the interactional forces between the stabilizing sheaths in steric stabilization. The elaboration of these effects has dominated recent developments in polymer solution thermodynamics, especially in the Prigogine-Flory free volume theory. These ideas can be directly transferred to illuminate the discussion of steric stabilization.

The free volume theories of polymer solution thermodynamics imply that at least three effects must be recognized as generating the interactional forces in steric stabilization. These are:

(i) combinatorial effects arising from the entropy of mixing of polymer segments with molecules of the dispersion medium. They always oppose flocculation, i.e., the associated free energy is always positive and increases with increasing temperature ($\Delta G_{comb} = -T\Delta S_{comb}$). They can be calculated using the classical Flory-Huggins expression for the entropy of mixing of polymer segments with solvent.

(ii) free volume (sometimes termed 'equation-of-state') contributions arising from the dissimilarities in the free volumes of the articulated macromolecules and the minimolecules of the dispersion medium. The corresponding free energy may contain both entropic and enthalpic contributions, although in nonaqueous media the entropic contributions are usually dominant and promote flocculation. These effects normally increase in absolute magnitude with increasing temperature because the free volume dissimilarity must increase as the temperature increases. In accordance with the adage 'like dissolves like', the free volume dissimilarity promotes flocculation.

(iii) contact dissimilarity contributions, which may be enthalpic and/or entropic in character. In nonaqueous media, the entropic effect is small or even non-existent and the dominant enthalpic effect usually promotes flocculation. The enthalpic effect would be expected to be relatively insensitive to temperature over small temperature ranges but might be expected to decrease in absolute magnitude with increasing temperature. In aqueous media, specific interactions between the water molecules and the stabilizing moieties may lead to enthalpic interactions that oppose flocculations, the concomitant entropic contributions promoting flocculation. This behaviour, whilst probably applicable to poly(oxyethylene) and poly(vinyl alcohol), is not universal. It is unlikely to be relevant to poly(acrylic acid) or poly(acrylamide). Unfortunately, because of the specific interactions that occur between polymer segments and water, there is as yet no adequate *ab initio* theory for the thermodynamics of aqueous polymer solutions. For this reason, some of the terms included in Table 4 for the free energy of close approach of sterically stabilized dispersions are speculative.

Table 4

Contributions to the Steric Interactional Free Energy
Nonaqueous (and Some Aqueous) Dispersions

	Combinatorial		Free Volume		Contact	
	ΔH	ΔS	ΔH	ΔS	ΔH	ΔS
	0	-ve	+ve (small)	+ve	-ve	~0
ΔG	+ve		-ve		-ve	
Promotes	stabilization		flocculation		flocculation	
$\partial(\Delta G)/\partial T$	+ve		-ve		+ve (small)	
Other Aqueous Dispersions						
	0	-ve	+ve	+ve	+ve	+ve
ΔG	+ve		-ve		+ve	
Promotes	stabilization		flocculation		stabilization	
$\partial(\Delta G)/\partial T$	+ve		-ve		-ve	

<u>Nonaqueous Dispersions</u>. The combinatorial entropy is usually responsible for stability under ambient conditions. On heating, the free volume dissimilarity contribution changes faster than the combinatorial term, resulting in flocculation. Just below the

ucft, $\Delta H_{free\ vol}$ gives rise to enthalpic stabilization. On cooling, the combinatorial term decreases and the contact dissimilarity term becomes relatively more important. Both the free volume and contact dissimilarity terms may contribute to flocculation on cooling. Just above the lcft, ΔS_{comb} gives rise to entropic stabilization. Of course, if the polymer is chemically similar to the dispersion medium (e.g., polyisobutylene in alkanes), the contact dissimilarity term, which is often dominant in promoting flocculation on cooling, is drastically reduced and may even be near zero. In these circumstances, the lcft may not be observable or, conceivably, may not even exist.

<u>Some Aqueous Dispersions</u>. Just below the ucft, $\Delta H_{contact}$ gives rise to enthalpic stabilization. On heating, the free volume contribution becomes more negative and the contact term less positive, resulting in flocculation. The lcft is often inaccessible in these systems but if it were observable, ΔS_{comb} would impart entropic stabilization just above the lcft. On heating these aqueous dispersions, the free volume contribution to the free energy becomes more negative and the contact term less positive, resulting in flocculation. The situation on cooling is uncertain: inspection of the entries in Table 4 shows that, unless some of the terms change sign on cooling or an important effect is missing, only the free volume entropy term promotes flocculation. This apparent absence of a strong flocculating mechanism may account for the experimental inaccessibility of the lcft in some aqueous systems. It is even possible that for such dispersions no lcft may exist. Like the nonaqueous case, however, our understanding of the polymer solution thermodynamics is still too primitive to confirm or exclude this possibility. Poly(oxyethylene) and poly(vinyl alcohol) would be typical of stabilizers that evince the pattern of behaviour described above.

Note that the entries in Table 4 relating to aqueous dispersions are far from exhaustive. Thus it is possible with biopolymers, e.g., for the contact entropy term to be positive due to the occurrence of so-called hydrophobic interactions, which would promote flocculation. It is not beyond the realm of possibility that hydrophobic interactions could be important with some synthetic polymers (e.g., poly(methacrylic acid)).

<u>Quantitative Prediction of the Repulsive Potential Energy</u>. It is
still not possible to make accurate *ab initio* predictions of the
repulsive potential energy in sterically stabilized systems in the
same sense as it is possible to make quantitative predictions
using the DLVO theory in some (although not all) electrostatically
stabilized systems. Two central problems remain unresolved:
prediction of polymer conformations at an interface; and the
thermodynamics of the interactions of the two steric barriers.
Of course, both are merely extensions of well-known problems in
classical polymer science, problems that have now defied exact
solution for some three decades. This is scarcely surprising
since they are both many-bodied problems. A diversity of
theoretical devices have been elaborated, however, to cope with
these problems in free solution and these theories have been
extended to the prediction of the repulsive potential energy in
steric stabilization.

<u>The Conformational Problem</u>. A wide range of models have been used
to simulate the conformational properties of polymers. These
range from the simple random flight (i.e., freely jointed) through
such models as the wormlike chain to the relatively complex
rotational isometric state model.[30] The latter model, relevant to
θ-solvents or the bulk phase, is perhaps the only one that can lay
claims to realistic quantitative predictions of conformational
properties. Other models, often of seductive simplicity, disregard
completely the essential features that distinguish one polymer
chain from another: specifically, they ignore the characteristic
interdependence of the rotational potentials of the bonds in
polymer chains and often take no specific account of the fixed
valency angles and bond lengths, which may vary from one bond to
another along the polymer chain (e.g., poly(oxyethylene)). Such
models are, however, qualitatively useful. It may even be possible
in the future to scale such models to give good agreement with
exact theories, although each conformational dependent property
may well have to be scaled differently. In any event, without an
exact model, such primitive models are prone to quantitative
impotence.

The conformational properties of polymers in solvents that
are better than θ-solvents, which are relevant to steric stabilization, are still not accessible by any exact theory, although an

attempt has been made by Mattice[31] to include the excluded volume effect into the rotational isomeric state formalism. At a simpler level, the self-avoiding chain is sometimes studied.[32-34] This presumably represents an extreme case of excluded volume and so can be used to delineate the broad features of the effects of excluded volume. But it is of limited quantitative value because it disregards an essential feature of polymer excluded volume, *viz.* that the excluded volume is a continuously variable function (as exemplified by the intramolecular expansion parameter α) rather than an all-or-nothing quantity, as assumed in the self-avoiding chain.

The presence of the interface simply complicates an unsolved problem even further. If the polymer interacts specifically with the surface so as to adsorb at various points along the chain, then the quantitative prediction of the conformation of the polymer becomes doubly difficult. Although it may be possible to incorporate a segmental adsorption free energy into the conformational theory, it is usually difficult to relate this quantity to that relevant to an actual system.

The conformation of isolated terminally attached polymer chains ('tails') at a noninteracting interface can be calculated using the rotational isomeric state procedure for the θ-solvent or bulk case by exploiting Monte Carlo procedures.[35, 36] These calculations demonstrate that the polymer chain in this case is elongated somewhat in a direction normal to the interface, the components parallel to the interface being unaffected. Whether the lateral approach of other chains on the surface of actual sterically stabilized particles results in additional extension normal to the interface is still a matter for conjecture.

The Solution Thermodynamics Problem

The deficiencies of the Flory-Huggins theory, in which most of the theories of steric stabilization are couched, are well documented.[37] These include its inability to account for a number of phenomena that typify nonaqueous polymer solutions: phase separation on heating; a strong dependence of the measured interaction parameter χ_1 on concentration; and an unexpectedly large entropic contribution to the interaction parameter and an unexpectedly low value for the enthalpic contribution.

It is possible, however, to use the Flory-Huggins theory in a purely pragmatic fashion. Experimental values for χ_1, including any concentration dependence,[38] can be used in the Flory-Huggins expressions for steric stabilization, whether the dispersion medium is aqueous or nonaqueous. The experimentally observed values of χ_1 can then be explained using the Prigogine-Flory free volume theory, which is applicable to many nonaqueous systems. Aqueous solutions still defy the elaboration of an *ab initio* theory because of the strong specific interactions inherent therein.

The Domains of Close Approach. When two polymer coated latex particles approach one another, no interaction is possible if the minimum distance between the particles exceeds 2L, where L = thickness of the steric stabilizing barrier. If the particles approach one another, so that the minimum distance lies between L and 2L, the stabilizing layers may interpenetrate. This interpenetration leads to repulsion in good solvents and to attraction in worse than θ-solvents. In the interpenetrational domain, the interactional free energy of close approach ΔG_T arises primarily from the mixing (or rather demixing) of polymer segments and molecules of the dispersion medium:

$$\Delta G_T = \Delta G_{mix}$$

When the distance of close approach decreases to less than L, nót only does interpentration occur but in the stabilizing layer on one particle may be compressed by the impenetrable surface of the other particle. This compression generates an elastic contribution to stabilization of purely entropic origin which always opposes flocculation:

$$\Delta G_T = \Delta G_{mix} + \Delta G_{elastic}$$

These two contributions do not necessarily have to be calculated separately; indeed, it may be argued[7] that this separation is artificial because the mixing and elastic terms are coupled. Nevertheless, the separation is physically useful in the sense that it provides a better insight into the origin of the repulsion generated on very close approach. Thus $\Delta G_{elastic}$ is still positive in worse than θ-solvents and provides a repulsive interaction close-in that has recently been measured directly under these conditions (see below).

Other descriptors[39, 40] have been used to designate the terms denoted here by 'mixing' and 'elastic': e.g., 'osmotic' and 'volume restriction'. It should be noted, however, that the designation 'osmotic' could be applied to almost all types of colloid stabilization, including electrostatic and depletion stabilization. The term 'osmotic' signifies a difference in chemical potential between the solvent in the interaction zone (located between the particles) and the bulk solution and most, although not all, types of stabilization depend upon this difference in the solvent chemical potential. Hence its use is inappropriate in this context. The designation 'elastic' is the one chosen by Flory to describe the change in configurational entropy due to conformational changes of chains in free solution.

The flocculation behaviour of sterically stabilized dispersions can usually be understood in terms of the interpenetrational domain. One notable exception is the flocculation of dispersions in a melt of the stabilizing polymer (see below).[41, 42] In that case, ΔG_{mix} is zero in the interpenetrational domain, and the elastic contribution in the interpenetrational + compressional domain must be invoked. The latter domain will also be important for understanding the compression of stable latices, or macroscopic plates, coated by polymer.

<u>Theories of Steric Stabilization</u>. Theories of steric stabilization can be classified according to whether or not they are *ab initio* theories, i.e., whether or not they aim at predicting the polymer conformation at the interface.

Ab Initio Theories. Meier[39] was perhaps the first to recognize clearly the existence of elastic and mixing contributions to steric stabilization and to exploit the diffusion equation to obtain an estimate of the magnitudes of these components. Subsequently, Hesselink, Vrij and Overbeek[40] refined these calculations via a statistical mechanical approach. Scheutjens and Fleer[43, 44] have latterly presented a more sophisticated statistical mechanical theory that has features reminiscent of the rotational isomeric state theory for conformations of polymers.

It must be stressed that all these theories use the freely jointed chain (random flight) model for the stabilizing moieties and its deficiencies have been noted above. Of course, in the limit of high molecular weight, all reasonably flexible polymers can be represented by a freely jointed chain, although a scaling factor for the number of bonds must be determined if the theory is to be applied to real chains.

Dolan and Edwards[45, 46] have also employed a random flight model in their treatment of the effects of excluded volume on steric stabilization. The excluded volume contributions were introduced via a self-consistent field approach. This treats the conformation as a random flight in the presence of a potential field that itself depends upon the solution of the random flight. Numerical iteration provides a self-consistent solution to the modified diffusion equation.

Note that in all the foregoing theories, the mixing contribution is incorporated using the Flory-Huggins theory. Note, too, that in the Dolan and Edwards approach, all contributions to the steric free energy are calculated concomitantly, the somewhat artificial separation into mixing and elastic contributions being circumvented. Their approach has been further elaborated by Levine *et al.*[47] and Gerber and Moore[48] to yield analytical solutions but it still suffers from the deficiencies of all random flight models.

Finally, we note that de Gennes[49, 50] has employed scaling theory to calculate the exponents of the distance dependence and molecular weight dependence relevant to steric stabilization.

<u>Pragmatic Theories</u>. These theories[7, 38] recognize the difficulties currently associated with predicting the conformations of adsorbed or attached chains at an interface. They assume that the segment density distribution functions can be determined experimentally (e.g., by neutron scattering) and simply calculate the mixing and elastic contributions using Flory-Huggins theory.[7, 38] Fortunately, the repulsive free energy is not dramatically sensitive to the precise details of the form adopted for the segment density distribution function. The concentration dependence of the interaction parameter can be included in these calculations if necessary.[38]

The most important result to emerge from such calculations is that the potential energy diagram for sterically stabilized

systems is radically different from that for electrostatically
stabilized systems. Whereas electrostatically stabilized
dispersions are at best thermodynamically metastable, sterically
stabilized systems may be thermodynamically stable. Figure 4
displays the expected form of the potential energy curves for
dispersion media that are either better solvents for the polymer
chains or worse solvents than θ-solvents.

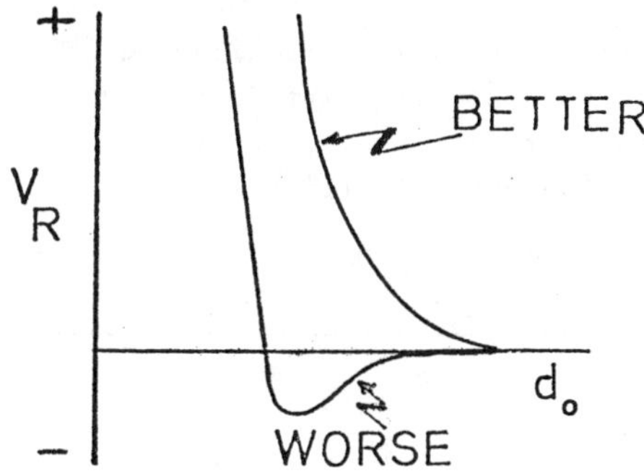

<u>Fig. 4</u> Schematic representation of the potential energy curves
for both better than and worse than θ-solvents.

<u>Experimental Measurements of Steric Interaction</u>. Klein[51] has
confirmed the general shape of one of the potential energy curves
shown in Fig. 4 by direct measurement of the forces of interaction
between two macroscopic mica plates coated by polystyrene in
cyclohexane below the θ-temperature. The general shape of the
curve in the case of a better than θ-solvent has also been
verified by Ottewill and coworkers.[52] They measured the increase
in the osmotic pressure of concentrated sterically stabilized
polymer latices on further compression. Note that considerable
progress has been made in simulating such dispersions by Monte
Carlo and molecular dynamics calculations.[53] An analytical
approach to calculating the pressure in colloidal systems has
also been reported.[54]

Enhanced Steric Stabilization. If the steric stabilizing moieties
are poorly anchored to the latex particles, the correlation
between the cfpt and θ-point is usually destroyed. Flocculation
results from the lateral movement and/or desorption of the
stabilizers under the stress generated by a Brownian collision.
Commonly, such flocculation occurs in better than θ-solvents.
Flocculation in better than θ-solvents also occurs if the
stabilizing moieties undergo crystallization from the dispersion
medium.[27, 55, 56]

The correlation between the cfpt and θ-point is also
weakened if the stabilizing chains are attached at a large number
of points along the chains (multipoint anchoring). Stability can
then be observed, however, in significantly worse than θ-solvents
(see Fig. 5). This phenomenon is termed 'enhanced steric stabilization'.[57]

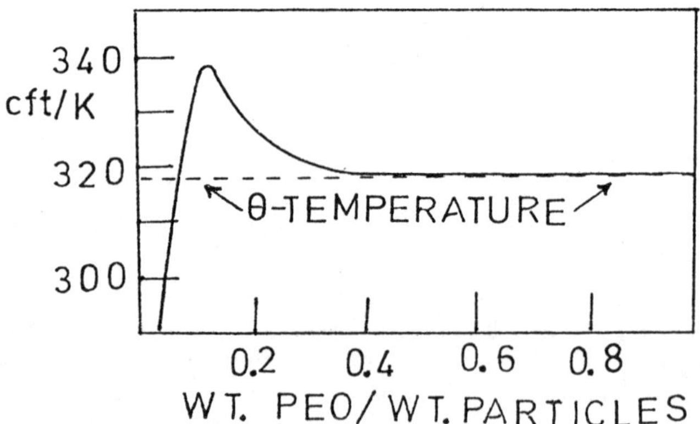

Fig. 5 Enhanced steric stabilization for poly(oxyethylene) chains
attached to polystyrene latex particles containing surface
carboxylic acid groups.

Random copolymers would be expected to impart enhanced
steric stabilization if one of the comonomers produces a homopolymer that is soluble in the dispersion medium whereas the
other comonomer generates an insoluble homopolymer.[24, 58]

The origins of enhanced steric stabilization are still obscure, although several possibilities have been proposed.[58] These include very low segment densities in the peripheral regions of the stabilizing sheath, changes in the thermodynamic parameters concomitant with the constraints imposed by multipoint anchoring, and the increased segment density close-in to the surface.

Elastic Steric Stabilization in Polymer Melts. Sterically stabilized dispersion can be prepared with polymer melts as the dispersion medium.[41, 42] This is important industrially in the dispersion of carbon black particles in rubber, in domain formation in melts of block or graft copolymers, and in some polymer composites. If, e.g., polystyrene latex particles are stabilized by poly(oxyethylene) of molecular weight 6000 in a melt of the same polymer, the mixing contribution to the free energy of close approach in the interpenetration zone must clearly be rendered negligible. Stability in this instance must arise from other sources. The most obvious origin is the elastic contribution resulting from the compression of the stabilizing sheath on one particle by the impenetrable surface of the other. Depletion effects, however, cannot be ruled out (see below) as contributing to stability.

The stability limit in polymer melts was found to depend upon the colloid particle size and the molecular weight of the stabilizer.[41, 42] Neither of these factors was found to influence the stability limit for sterically stabilized particles in a minimolecular dispersion medium.

Heterosteric Stabilization. Dispersed particles stabilized by different types of stabilizing moieties may exhibit heteroflocculation or heterosteric stabilization on mixing. Thus, e.g., it is possible to mix two aqueous homosterically stabilized dispersions, one homosterically stabilized by poly(oxyethylene) and the other by poly(acrylamide), without the occurrence of flocculation. On the other hand, at suitably low pH values, the mixing of two aqueous homosterically stabilized dispersions, one stabilized by poly(oxyethylene) and the other by poly(acrylic acid), can result in heteroflocculation.

The theory for heterosteric stabilization has been developed,[59] both for minimolecular and macromolecular dispersion media. It predicts that stability will normally be observed for low molecular weight dispersion media on mixing particles stabilized by incompatible polymers whereas compatible polymers, especially those that coprecipitate or coacervate, will often undergo heteroflocculation on mixing. Croucher and Hair[17] have tested the theoretical predictions of Feigin and Napper[59] experimentally by mixing dispersions that were separately homosterically stabilized in cyclopentane by two incompatible polymers: polystyrene and polyisobutylene. As predicted, such mixed dispersions display heterosteric stabilization and not heteroflocculation.

Another important conclusion to emerge from the theoretical analysis was the possibility of the selective flocculation of mixed sterically stabilized dispersions. Croucher and Hair[17] have verified the practicability of selective flocculation in nonaqueous media. The mixture of dispersions mentioned above heterosterically stabilized by polystyrene and polyisobutylene in cyclopentane was found to flocculate selectively on changing the temperature to the lcft or the ucft of the dispersion stabilized by polystyrene. In both instances, the residual particles flocculated at the ucft of the dispersion stabilized by polyisobutylene.

<u>Stabilization by Free Polymer: Depletion Stabilization</u>

Asakura and Oosawa[60] were perhaps the first to recognize in 1958 that free polymer when added at relatively low concentrations to the dispersion medium could induce flocculation. This was subsequently investigated, either theoretically and/or experimentally, by Sieglaff,[61] Vrij,[62] de Gennes[63] and Vincent *et al*.[64,65]

The origin of this phenomenon, termed 'depletion flocculation', is straight forward.[66,67] It arises whenever the colloidal particles are so close as to exclude polymer chains from the interparticle region (see Fig. 6). Crudely, this exclusion occurs when the interparticle distance is less than the width of the polymer molecule (which for linear polymers is roughly the r.m.s. end-to-end distance of the chains); closer approach of the particles is then favoured by the lowering in free energy resulting from the mixing of (almost) pure solvent with bulk polymer solution, obviously a spontaneous process.

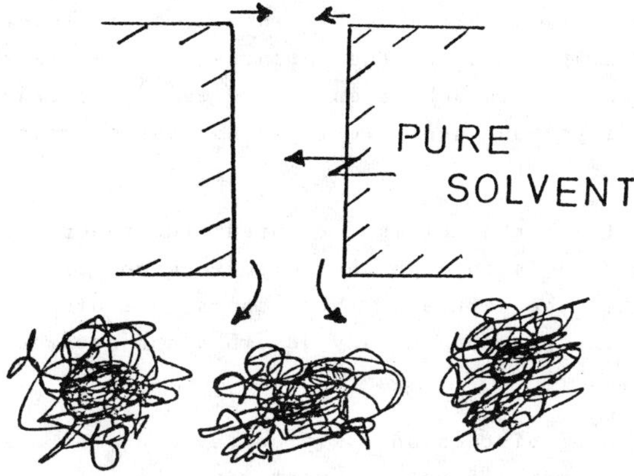

<u>Fig. 6</u> The origin of depletion flocculation

In good solvents, the polymer depleted regions between
the particles can only be obtained by demixing polymer chains and
solvent, a process that is thermodynamically disfavoured. Consequently, work must be done in order to bring the particles to an
interparticle distance equal to the width of the polymer chains.
This corresponds to the existence of a repulsive potential energy
between the particles, a repulsion that can be sufficient to give
rise to stability at higher concentrations of polymer than that
required to induce flocculation. This is termed 'depletion
stabilization'.

The foregoing discussion implies that the potential energy
diagram for depletion stabilization possesses a maximum at an
interparticle separation roughly comparable to that of the width of
the free polymer (see Fig. 7). In this respect, depletion
stabilization more closely resembles electrostatic stabilization
than it does steric stabilization in that such systems are thermodynamically metastable rather than displaying the thermodynamic
stability of sterically stabilized systems.

The existence of this maximum is readily seen from Fig. 6 if
the interparticle distance is comparable to the width of the free
polymer. Closer approach, as noted above, causes the pure solvent
to mix with the bulk solution, thus lowering the free energy of

the system; but equally, if the particles are moved apart, the
pure solvent can mix with the bulk solution, again lowering the
free energy of system. Hence the potential energy diagram must
display a maximum. Both Scheutjens and Fleer[68] and Clark and
Lal[69] have predicted on theoretical grounds the existence of this
maximum.

The effects of the polymer are predicted theoretically[67]
to depend upon such factors as the polymer composition and
molecular weight, the nature of the dispersion medium, the particle size and the presence of any adsorbed and attached polymer
at the interface.

The foregoing discussion was predicated on the assumption
that the free polymer did not interact with the particles. It
would usually be expected that free polymer would adsorb onto the
particle surfaces. The presence of such steric layers significantly complicates any theoretical treatment of depletion
stabilization because they tend to counteract the effects of the
depletion zones adjacent particle surfaces. This induces a non-
additivity in the potential energies associated with depletion
and steric effects due to their cross-coupling, the precise
details of which still have to be elaborated.

Experimentally, however, the presence of thin steric layers
permits the phenomenology of depletion flocculation to be more
readily discerned by eliminating the effects of van der Waals
interaction. Vincent *et al.*[64, 65] have shown that systems
stabilized sterically by low molecular weight poly(oxyethylene)
can be flocculated by the addition of free poly(oxyethylene) of
much higher molecular weight. Flocculation is only observed in
a restricted domain of concentration of free polymer, which can
be predicted theoretically using the concepts set forth above.[66,67]

The restricted flocculation domain can be readily understood.
At low polymer concentration, the minimum in the free energy curve
(see Fig. 7) caused by entry into the close approach domain
depicted in Fig. 6 is insufficient to induce flocculation and the
system is primarily sterically stabilized. As the concentration
of free polymer is increased so the minimum becomes deeper,
resulting at some critical concentration in depletion flocculation.
Note that a potential energy minimum of only a few kT is sufficient

for flocculation to be observed and so this is apparent at low
polymer concentrations.

At higher concentrations, the maximum in the potential energy
curve becomes sufficiently large to prevent the particles from
entering the minimum that would be sufficient to induce flocculation. A potential energy maximum of order 20 kT would be required
for stability to be observed. This is significantly greater than
the absolute value required for the depth of the potential energy
minimum if the latter is to induce flocculation. Thus depletion
stabilization is only observed at polymer concentration in excess
of that required to cause the onset of depletion flocculation.

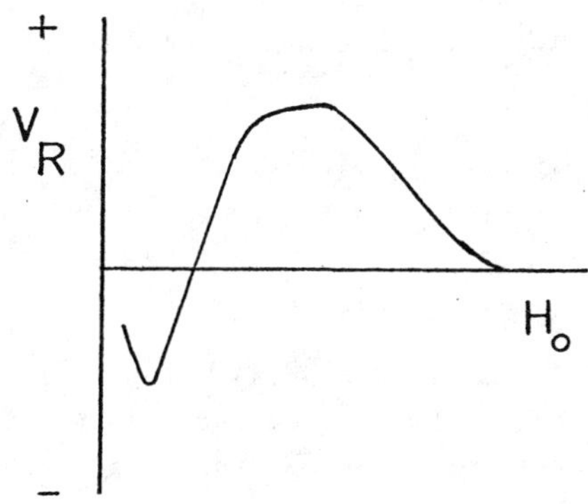

Fig. 7 A potential energy diagram for depletion stabilization in
the presence of thin steric barriers.

Schematic Representation of the Effects of Idealized High
Molecular Weight Polymer

Fleer[70] has devised the scheme shown in Fig. 8 for the
effects of idealized polymers on colloid stability.

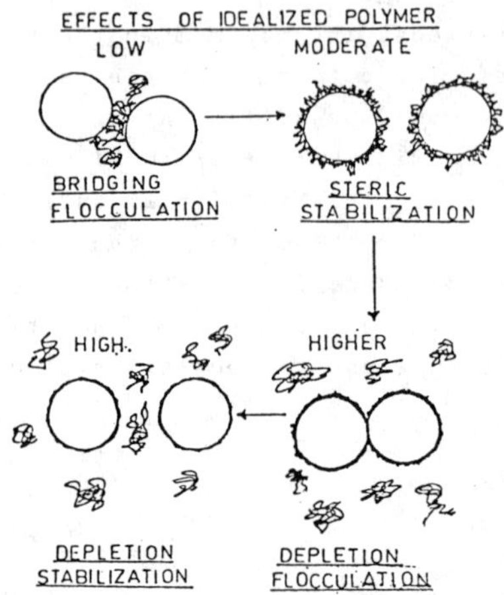

<u>Fig. 8</u> Schematic representation of the effects of idealized
polymers on colloid stability.

Acknowledgements

I thank the ARGC for its financial support of these
studies and Dr. R. Evans, J.B. Smitham and R.I. Feigin for
their many contributions to my understanding of the effects
of polymers.

References

[1] <u>Encyclopaedia Brittanica</u>, 1968, <u>12</u>, 257; <u>17</u>, 38.

[2] J. Mahanty and B.W. Ninham, "Dispersion Forces", Academic,
London, 1976.

[3] B. Vincent, <u>Adv. Colloid Interface Sci.</u>, 1974, <u>4</u>, 193.

[4] Th.F. Tadros, <u>Adv. Colloid Interface Sci</u>., 1980, <u>12</u>, 141.

[5] G.D. Parfitt and J. Peacock, in "Surface and Colloid Science",
ed. E. Matijevic, Wiley, New York, 1979, chap. 4.

6 T. Sato and R. Ruch, "Stabilization of Colloidal Dispersions by Polymer Adsorption", Dekker, New York, 1980.

7 D.H. Napper, J. Colloid Interface Sci., 1977, 58, 390.

8 K.E.J. Barrett, "Dispersion Polymerization in Organic Media", Wiley, London, 1975.

9 W. Heller, Pure Appl. Chem., 1966, 12, 249.

10 D.H. Napper, J. Colloid Interface Sci., 1970, 32, 106.

11 D.H. Napper, Trans. Faraday Soc., 1968, 64, 1701.

12 G.C. March and D.H. Napper, J. Colloid Interface Sci., 1977, 61, 383.

13 M.D. Croucher and M.L. Hair, J. Colloid Interface Sci., to be published.

14 M.D. Croucher and M.L. Hair, Macromolecules, 1978, 11, 874.

15 M.D. Croucher and M.L. Hair, J. Phys. Chem., 1979, 83, 1712.

16 M.D. Croucher and M.L. Hair, in "Polymer Colloids II", ed. R.M. Fitch, Plenum, New York, 1980, p. 497.

17 M.D. Croucher and M.L. Hair, Colloids Surfaces, 1980, 1, 349.

18 J.V. Dawkins and G. Taylor, Colloid Polym. Sci., 1980, 258, 79.

19 R. Evans, J.B. Davison and D.H. Napper, Polym. Letts., 1972, 10, 449.

20 R. Evans, D.H. Napper and A.H. Ewald, J. Colloid Interface Sci., 1975, 51, 552.

21 R. Evans and D.H. Napper, J. Colloid Interface Sci., 1975, 52, 260.

22 R. Evans, B.Sc. Hons. Thesis, University of Sydney, 1971.

23 B. Vincent, Faraday Discuss. Chem. Soc., 1978, 65, 313.

24 R.J.R. Cairns and E.L. Neustadter, Proc. Int. Conf. Colloid Surface Sci., 1975, 1, 323.

25 E.L. Neustadter, Faraday Discuss. Chem. Soc., 1978, 65, 318.

26 D.H. Everett and J.F. Stageman, Colloid Polym. Sci., 1977, 255, 293.

27 D.H. Everett and J.F. Stageman, Faraday Discuss. Chem. Soc., 1978, 65, 231.

28 R.I. Feigin, J. Dodd and D.H. Napper, Colloid Polym. Sci., accepted for publication, 1981.

29 P.J. Flory, Discuss. Faraday Soc., 1970, 49, 7.

30 P.J. Flory, "Statistical Mechanics of Chain Molecules", Interscience, New York, 1969.

31 W.L. Mattice and G. Santiago, Macromolecules, 1981, to be published.

32 F.T. Wall, W.A. Seitz, J.C. Chin and P.G. de Gennes, Proc. Natl. Acad. Sci. USA, 1978, 75, 2069.

33 A.T. Clark and M. Lal, J. Chem. Soc. Faraday Trans. II, 1980, accepted for publication.

34 A.J. Guttmann, K.M. Middlemiss, G.M. Torrie and S.G. Whittington, J. Chem. Phys., 1978, 69, 5375.

35 R.I. Feigin and D.H. Napper, J. Colloid Interface Sci., 1979, 71, 117.

36 W.L. Mattice and D.H. Napper, Macromolecules, accepted for publication, 1981.

37 E.F. Casassa, J. Polym. Sci. C, 1976, 54, 53.

38 R. Evans and D.H. Napper, J. Chem. Soc. Faraday Trans. I, 1977, 73, 1377.

39 D.J. Meier, J. Phys. Chem., 1967, 71, 1861.

40 F.Th. Hesselink, A. Vrij and J.Th.G. Overbeek, J. Phys. Chem., 1971, 75, 2094.

41 J.B. Smitham and D.H. Napper, J. Colloid Interface Sci., 1976, 54, 467.

42 J.B. Smitham and D.H. Napper, J. Chem. Soc. Faraday Trans. I,
1976, 72, 2425.

43 J.M.H.M. Scheutjens and G.J. Fleer, J. Phys. Chem., 1979,
83, 1619.

44 J.M.H.M. Scheutjens and G.J. Fleer, J. Phys. Chem., 1980,
84, 178.

45 A.K. Dolan and S.F. Edwards, Proc. Roy. Soc. London Ser. A.,
1974, 337, 509.

46 A.K. Dolan and S.F. Edwards, Proc. Roy. Soc. London Ser. A.,
1975, 343, 427.

47 S. Levine, M.M. Thomlinson and K. Robinson, Faraday Discuss.
Chem. Soc., 1978, 65, 202.

48 P.R. Gerber and M.A. Moore, Macromolecules, 1977, 10, 476.

49 P.G. de Gennes, "Scaling Concepts in Polymer Physics",
Cornell, Ithaca, 1979.

50 P.G. de Gennes, private communication, 1980.

51 J. Klein, Nature, 1980, 288, 248.

52 R.J.R. Cairns, R.H. Ottewill, D.W.J. Osmond and I. Wagstaff,
J. Colloid Interface Sci., 1976, 54, 45.

53 R.J.R. Cairns, W. van Megen and R.H. Ottewill, J. Colloid
Interface Sci., 1981, 79, 511.

54 R. Evans and D.H. Napper, J. Colloid Interface Sci., 1978,
63, 43.

55 M.D. Croucher, M.L. Hair and P.R. Sundararajan, Colloid
Polym. Sci., 1980, 258, 408.

56 P.R. Sundararajan, G.K. Hamer and M.D. Croucher, Macromolecules,
1980, 13, 971.

57 J.W. Dobbie, R. Evans, D.V. Gibson, J.B. Smitham and D.H. Napper,
J. Colloid Interface Sci., 1973, 45, 557.

58 J.B. Smitham and D.H. Napper, Colloid Polym. Sci., 1979, 257, 748.

59 R.I. Feigin and D.H. Napper, J. Colloid Interface Sci., 1978, 67, 127.

60 S. Asakura and F. Oosawa, J. Polym. Sci., 1958, 33, 183.

61 C.L. Sieglaff, J. Polym. Sci., 1959, 41, 319.

62 A. Vrij, Pure Appl. Chem., 1976, 48, 471.

63 J.F. Joanny, L. Liebler and P.G. de Gennes, J. Poly. Sci. Polym. Phys. Ed., 1979, 17, 1073.

64 F.K.R. Li-in-on, B. Vincent and F.A. Waite, A.C.S. Symp Ser., 1975, 9, 165.

65 C. Cowell, F.K.R. Li-in-on and B. Vincent, J. Chem. Soc. Faraday Trans I, 1978, 74, 337.

66 R.I. Feigin and D.H. Napper, J. Colloid Interface Sci., 1980, 74, 567.

67 R.I. Feigin and D.H. Napper, J. Colloid Interface Sci., 1980, 75, 525.

68 J.M.H.M. Scheutjens and G. Fleer, private communication, 1981.

69 A.T. Clark and M. Lal, J. Chem. Soc. Faraday Trans II, accepted for publication, 1981.

70 G. Fleer, private communication, 1981.

6
Light Scattering

By P. N. Pusey

ROYAL SIGNALS AND RADAR ESTABLISHMENT, MALVERN, WR14 3PS, U.K.

1 Introduction

Light scattering, as a probe of colloidal suspensions, has a long history. The theoretical development of the subject goes back more than a century with notable contributions from Maxwell, Tyndall, Rayleigh, Lorenz, Einstein, Smoluchowski, Mie, Debye and many others (see Kerker[1] for a review). Experimental development of conventional (or time-averaged) light scattering (CLS) started around 1940. Here one measures the absolute intensity of light scattered by the suspension, usually as a function of scattering angle. An important landmark in the progress of light scattering was the invention of the laser in 1961. Lasers not only provided intense, well-collimated light sources but, more importantly, led to the development of a totally new technique, dynamic light scattering (DLS). Dynamic light scattering exploits the coherence of laser light to provide information about the dynamics (i.e. the Brownian motions) of the particles in a colloidal suspension.

An excellent review of CLS is given by Kerker[1]; references 2 and 3 are also useful. Several good articles and books on DLS (though not usually slanted specifically towards colloid science) also exist[4-13].

The aim here is to give a very brief description of the principles and applications (in colloid science) of both CLS (section 2) and DLS (section 3). In section 4 we compare, as an illustration of light scattering experiments, recent determinations of particle size by CLS and DLS. Finally, in section 5, we mention some of the difficulties encountered in the practice of light scattering and compare briefly the uses of light scattering and neutron scattering in colloid science.

For simplicity, emphasis in this article will be on homogeneous

spherical particles although the principles outlined can be
generalized to treat other types of particle such as spheres of
variable composition, ellipsoids, rods, flexible polymers etc.

2 Conventional light scattering

2.1 <u>Introduction</u>. Associated with a beam of light incident on
a colloidal suspension is an electric field which oscillates at
frequency $\nu = c/\lambda$, where c is the velocity of light and λ its wavelength in the suspension. Provided there is a difference between
the refractive index n_1 of the particles and that n_2 of the liquid
in which they are suspended, this electric field induces in a
particle a dipole moment which also oscillates at frequency ν and
therefore causes a reradiation or scattering of the light in all
directions. In section 2 we consider first (sections 2.2 - 2.4)
the scattering by isolated spheres of various relative sizes R/λ
(where R is the particle radius), then (section 2.5) the scattering by a dilute suspension of identical particles and finally
(section 2.6) some effects of polydispersity (a distribution of
particle size).

We assume the electric vector of the incident light to be
polarized perpendicular to the scattering plane and observe the
scattered radiation with this same polarization at a detector set
at scattering angle θ at a distance r from the sample. (Further
details can be found in Kerker[1].)

2.2 <u>Rayleigh (point) scatterer, $R/\lambda \ll 1$</u>. The intensity I_R of
light scattered by an isolated particle of size much smaller than
λ is

$$I_R = \frac{16\pi^4 R^6}{r^2 \lambda^4} \left[\frac{n_1^2 - n_2^2}{n_1^2 + 2n_2^2} \right]^2 \tag{1}$$

where all quantities have been defined in section 2.1 and the sub-
script R indicates a "Rayleigh" or point scatterer. Note: (i) I_R
is independent of scattering angle; (ii) it depends on the refractive index difference $n_1 - n_2$ and becomes zero when $n_1 = n_2$; (iii) it
is proportional to R^6, since the scattered field amplitude $\sqrt{I_R}$
is proportional to the polarizability (or volume) of the sphere;
(iv) it goes as λ^{-4}, the well-known Rayleigh scattering law.

2.3 <u>Rayleigh-Gans-Debye (RGD) scatterer, $(n_1 - n_2)R/\lambda \ll 1$</u>. The
electric fields scattered (at non-zero θ) from different parts of
a particle whose size is comparable to λ suffer relative phase

shifts of order 2π. Thus interference between the different ele-
mental scattered fields occurs at the detector and the scattered
intensity is reduced relative to that of a Rayleigh particle. In
the Rayleigh-Gans-Debye limit

$$(n_1 - n_2)\ R/\lambda\ <<\ 1, \tag{2}$$

the incident light wave is not distorted significantly on passage
through the particle and

$$I_{RGD}\ =\ I_R\ x\ P(\theta), \tag{3}$$

where $P(\theta)$ is a shape factor with the properties $P(O) = 1$,
$P(\theta) < 1$ for $\theta > O$. For homogeneous spheres, $P(\theta)$ takes the
relatively simple form

$$P(\theta)\ =\ \left[3(\sin QR - QR\cos QR)/(QR)^3\right]^2 \tag{4}$$

where the "scattering vector" Q is given by

$$Q\ =\ (4\pi/\lambda)\ \sin(\theta/2). \tag{5}$$

Expansion of equation 4 in powers of QR gives

$$P(\theta)\ =\ 1 - (QR)^2/5 + ... \tag{6}$$

and it can be shown that, for RGD particles of arbitrary shape and
structure,

$$P(\theta)\ =\ 1 - (QR_G)^2/3 + ... \tag{7}$$

where R_G is the particle's radius of gyration ($R_G = \sqrt{3/5}\ R$ for a
homogeneous sphere).

2.4   Mie scatterer, $(n_1-n_2)R/\lambda \gtrsim 1$. When the particle is large
enough that the RGD criterion (equation 2) is not fulfilled, the
incident light wave can be severely distorted on passing through
the particle and the theoretical situation is much more complicated. For very large particles, $R >> \lambda$, an exceedingly complex
angular dependence of the scattered intensity is found which can
reasonably be described by ray optics i.e. in terms of rays reflected, both externally and internally (many times), and refrac-
ted by the particle. The angle-dependence of the intensity is then
a very sensitive measure of particle size and structure (see
section 4 for an example).

2.5   Dilute suspension of identical RGD particles. At sufficiently low concentrations, the positions of particles in a suspension are essentially uncorrelated. Then, if a conventional
(incoherent) light source is used, the intensities scattered by

different particles simply add at the detector. With a laser
(coherent) source, temporal fluctuations are observed in the
scattered intensity (see section 3); however, if this intensity
is averaged over many fluctuation times, it is again simply the
sum of the intensities scattered by the individual particles.
Then the intensity scattered by a suspension of N indentical particles at angle θ is, from equations 1 and 3,

$$I(\theta) \; = \; \frac{16\pi^4 \, R^6 \, N}{r^2 \lambda^4} \left(\frac{n_1^2 - n_2^2}{n_1^2 + 2n_2^2} \right)^2 P(\theta) \tag{8}$$

which can be rewritten in the form

$$I(\theta) \; = \; \frac{9 \, \pi^2 \, V_S \, C \, M}{r^2 \lambda^4 \, N_A} \left[\frac{1}{\rho} \left(\frac{n_1^2 - n_2^2}{n_1^2 + 2n_2^2} \right) \right]^2 P(\theta), \tag{9}$$

where V_S is the "scattering volume", C, the concentration of particles by weight, M, the particle molecular weight, ρ, the particle
density and N_A, Avogadro's number.

Thus measurement, on an absolute scale, of the intensity
(extrapolated to θ = 0, where P(θ) = 1) scattered by a suspension
of particles gives their molecular weight since all other quantities in equation 9 are, in principle, measurable. Even if absolute measurements are not made, the initial slope of a plot of
ℓn I(θ) against Q^2 gives, through equations 5, 7 and 9, an estimate of the particle's radius of gyration. In practice this
"Guinier" method of particle sizing is limited to radii of gyration in the approximate range 200 Å < R_G < 1000 Å; (for particles
with $R_G \lesssim 200$ Å the second term in equation 7 is small at all θ,
whereas for $R_G \gtrsim 1000$ Å higher terms are important even at small
angles). The quantity in square brackets in equation 9 can be
related to the refractive index increment dn/dC of the suspension[1];
then equation 9 becomes applicable to RGD particles of arbitrary
shape and structure.

2.6   <u>The effects of polydispersity</u>. In colloid science, in particular, it is rare for all the particles in a suspension to have
essentially the same size. It is thus important to evaluate the
effects of polydispersity. A simple generalization of equations
6 and 8 shows that, for homogeneous spheres, the apparent radius
obtained from an absolute intensity measurement (at θ → 0) is
given by

$$R_{app}^3 \; = \; \int F(R) \, R^6 \, dR \Big/ \int F(R) \, R^3 \, dR \tag{10}$$

whereas the apparent radius obtained from the angle-dependence is
given by

$$R^2_{app} = \int F(R)\, R^8\, dR \Big/ \int F(R)\, R^6\, dR \tag{11}$$

where $F(R)$ is the distribution of particle radii. The appearance
of high moments of the particle size distribution stems from the
R^6-dependence in equation 1 and is found also in dynamic light
scattering (see section 3). In the case of general RGD particles,
equations 10 and 11 can be written in terms of a molecular weight
distribution. Then absolute intensity measurements give a <u>weight-
average</u> molecular weight whereas the angle-dependence gives the
<u>Z-average</u> of R^2_G.

3 Dynamic light scattering

3.1 <u>Introduction</u>. As mentioned previously, the development of
dynamic light scattering followed soon after the invention of the
laser. For present purposes the important property of laser light
is that it is <u>coherent</u> so that phase relationships are maintained
in the scattering process. Thus the particles in a suspension can
be regarded as forming a random three-dimensional diffracting array
which gives rise to a random diffraction or "speckle" pattern consisting of small bright spots (where largely constructive interference occurs between the light fields scattered by the individual
particles) and dark areas (where destructive interference occurs).
Of course particles suspended in a liquid are not stationary but
are in perpetual erratic Brownian motion caused by collisions with
the thermally-agitated liquid molecules. As the particles move,
the phase relationships determining the speckle pattern change,
and the pattern itself changes continually through a series of
random configurations. Not surprisingly, the detailed nature of
the temporal fluctuations in the scattered intensity contains
information on the particle motions.

Two slight digressions are instructive here. Firstly there
is a close formal analogy between the speckle pattern formed by
laser light scattered by a particle suspension and the diffraction
pattern of X-rays scattered by a crystal. The main difference
is that the atoms in a crystal are constrained to the neighbourhood of sites on a regular lattice. Thus the diffraction pattern
consists of an ordered array of relatively few Laue spots whose
intensity hardly fluctuates. Secondly, when a particle suspension
is illuminated by light from a conventional (rather than laser)

source, due to its lack of coherence, phase relationships are only
maintained over a few inter-particle spacings rather than over the
whole scattering volume. The scattered intensity pattern then
consists of many independent speckle patterns superimposed so that
the fine spatial structure is smeared out (and the relative intensity fluctuations vastly reduced). Nevertheless by careful frequency and spatial filtering of a conventional source it is still
possible to observe intensity fluctuations in the scattered light.
This was suggested by Ramachandran[14] and Raman in 1943 (apparently
the first, albeit rudimentary, description of DLS) and verified
experimentally in 1976 by Jakeman et al[15].

Since about 1970, analysis of the fluctuating intensity in a
DLS experiment has usually been performed by the technique of
"photon correlation spectroscopy" (PCS). One fluctuating speckle
in the scattered intensity pattern illuminates a photomultiplier
tube operating in the photon-counting mode. The temporally-
modulated digital signal so obtained is fed to a purpose-built
digital computer (the "photon correlator") which, by a process
described in detail in the articles cited in section 1, determines
$g(\tau)$, the so-called temporal correlation function of the scattered
light field. This function, $g(\tau)$, is the analogue in time space
τ of the frequency or power spectrum of the intensity fluctuations;
some of its properties are discussed in the next sections.

3.2 <u>Dilute suspension of identical spheres</u>. The simplest non-
trivial application of DLS is to a suspension of identical spheres
at low enough concentrations that correlations between the positions
and motions of different particles can be neglected. It can then
be shown that $g(\tau)$, the quantity measured by DLS, takes the simple
form:

$$g(\tau) = \exp(-\tau/T_c) \tag{12}$$

where

$$T_c = 1/DQ^2, \tag{13}$$

the scattering vector Q is defined by equation 5 and τ is the
"correlation delay time"; the translational diffusion coefficient
D of a particle is given by the Stokes-Einstein relationship

$$D = kT/6\pi\eta\, R_H \tag{14}$$

where k is Boltzmann's constant, T the temperature, η the viscosity
of the liquid and R_H the "hydrodynamic radius" of the particle.

In equation 12 T_c can be interpreted as the typical fluctuation
time of the speckle pattern at the detector. From equations 5 and
13 it is seen that T_c is roughly the time taken by a particle to
diffuse a distance λ thereby changing significantly the phase relationships between the light scattered by different particles
and, in consequence, causing a significant change in the speckle
pattern. For submicron particles and typical values of Q, T_c is
in the range of microseconds to milliseconds.

In this simple case, a measurement of $g(\tau)$ by DLS provides,
through equation 12, the fluctuation time T_c. Since Q, T and η are
all measurable quantities, use of equations 13 and 14 then gives
the particle hydrodynamic radius R_H. Under favourable conditions R_H
can be determined with an accuracy of a few percent in an experiment of duration a minute or so. In many cases R_H is close to the
actual particle radius R (see section 4 for further discussion).
Dynamic light scattering has thus become a widely-accepted method
for sizing particles particularly in the range from about 20 $\overset{\circ}{A}$
to 1 μm where there are few competing techniques.

3.3 <u>Polydispersity</u>. For polydisperse non-interacting RGD spheres
equations 12 and 13 are easily generalized to

$$g(\tau) = \frac{\int F(R)\ I(R,\theta)\ \exp[-D(R)\ Q^2\tau]\,dR}{\int F(R)\ I(R,\theta)\,dR}\ , \qquad (15)$$

where F(R) is, as before, the distribution of particle radii,
$I(R,\theta)$ is given by equation 8 and D(R) by equation 14 (if we take
$R_H = R$). It is seen that $g(\tau)$ now consists of a sum of exponentials, one for each species of particle, weighted by the intensity
scattered by each species. Speaking mathematically, the measured
quantity $g(\tau)$ is the Laplace transform of the product F(R) $I(R,\theta)$.
In principle, therefore, it should be possible to invert equation
15 to obtain an estimate of F(R) $I(R,\theta)$ from a measurement of $g(\tau)$
which, with use of equation 8, would provide the particle size
distribution F(R). Unfortunately this is an "ill-conditioned"
problem which means that a small amount of statistical uncertainty
in the measurement of $g(\tau)$ is converted, in the inversion process,
to a large uncertainty in the desired quantity F(R). In simple
terms, this is just an expression of the well-known fact that a
sum of exponentials whose rates of decay are not too different
looks very like an appropriately defined <u>average single</u> exponential.
Nevertheless, for relatively broad particle size distributions,
various methods[12,16] for numerical inversion of equation 15 show

promise although they have yet to be fully proved and widely
accepted.

For narrower distributions it is possible to expand the exponential in equation 15 about a mean value to give

$$\ln g(\tau) \;=\; -AQ^2\tau \;+\; \frac{B}{2}\,Q^4\tau^2 \;+\; \ldots \tag{16}$$

where the coefficients A,B etc are related to moments of the particle size distribution. For spherical Rayleigh scatterers (assuming $R_H = R$),

$$A \;=\; \frac{kT}{6\pi\eta\,(\overline{R^6}/\overline{R^5})} \tag{17}$$

and

$$B/A^2 \;=\; \frac{\overline{R^6}\;\overline{R^4}}{\overline{R^5}^{\,2}} \;-\; 1, \tag{18}$$

where $\overline{R^n}$ is the n'th moment of F(R). By fitting the logarithm of
the measured correlation function $g(\tau)$ to equation 16, it is thus
possible to obtain a well-defined, if unusual, average radius and
a measure, B/A^2, of the width of the particle size distribution.

3.4   <u>Other applications of DLS</u>. For simplicity, in this short
article, we have emphasized what is, to date, probably the most
widely-used application of DLS in colloid science, the sizing of
spherical particles. However there are other applications of
growing importance, some of which are mentioned below.

With non-spherical particles comparable in size to λ the form
factor $P(\theta)$ becomes dependent on the orientation or configuration
of the particle; there are then fluctuations in the scattered

intensity associated with rotational and internal configurational
motions. These lead to extra terms in equation 12 and it is
possible to measure rotational diffusion constants of rigid non-
spherical particles as well as flexing motions of polymers. Similar information can be obtained from the study of intensity fluctuations in the depolarized scattered light (if this is sufficiently
intense).

If the particles are charged, strong correlations between the
positions and motions of different particles, caused by long-ranged
coulombic interactions, can occur. Similar effects are found in
concentrated suspensions of uncharged particles. CLS can then be
used to study the average spatial arrangement of particles in the

suspension, characterized by the structure factor and the radial distribution function (see the articles on Concentrated Dispersions and Neutron Scattering in this volume). The dynamics of interacting particles, which can be studied by DLS, form a complicated but important subject[17].

When charged particles experience an externally-applied electric field, a coherent motion is superimposed upon the random Brownian motion. By mixing laser light scattered by these particles with unscattered laser light it is possible to measure the Doppler shift and hence the drift velocity associated with this coherent motion[18]. Compared to conventional methods of electrophoresis, those based on DLS have the advantages of speed and accuracy and, in the case of small particles, of not requiring the creation of a macroscopic boundary.

Certain biological cells and micro-organisms such as spermatozoa exhibit motility, spontaneous self-propulsion or "swimming" which dominates natural Brownian motion. DLS can be used to investigate these motions[19] and is being developed towards a standard assay of motility in both animal and human fertility studies.

Finally DLS can be used to measure thermal diffusion and viscous relaxation in pure liquids as well as the dynamics of concentration fluctuations in mixtures of liquids. It has been particularly valuable in the study of dynamic critical phenomena in these systems[20].

4   Experimental data

Here we present some experimental data obtained by light scattering from a suspension of fairly monodisperse spheres of radius about 1.7 μm. (This system was recently studied in detail in order to investigate a predicted failure of equations 12 and 13 at short times $\tau \ll T_c$; the original paper[21] can be consulted for further details.) The data points in figure 1 show the angle dependence of the average intensity of light scattered by this suspension. The same set of data is plotted three times. Many theoretical Mie scattering profiles were generated by computer and three of these are shown in figure 1; α is the size parameter,

$$\alpha = 2\pi R/\lambda , \tag{19}$$

and $m \equiv n_1/n_2$ is the relative refractive index. The most noticeable difference between experiment and theory is that the data do

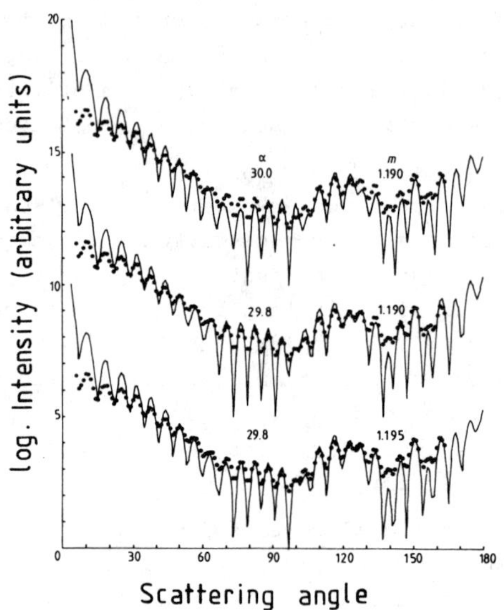

Figure 1

not show the predicted deep minima; this is probably caused by
multiple scattering and/or a small degree of polydispersity. Two
features of the theoretical curves were found to be particularly
sensitive to variations in α and m: the depth of the modulation
around θ = 50° and the "shoulder" at θ≈98°. Using these as
criteria we judged that the middle curve fits the data best.
Taking α = 29.8 ± 0.3 then gives, through equation 19, R = 1.69 ±
0.02 μm.

Figure 2 shows a plot of the logarithm of the correlation
function $g(\tau)$, measured by DLS, against delay time τ. As predicted by equation 12 the data fall on a straight line; (the
failure of equation 12, mentioned above, is expected to have only
a small effect on the first point or two of this plot). Analysis
of these data, with use of equations 5, 12, 13 and 14, gives
R_H = 1.75 ± 0.02 μm.

We thus have two independent measurements of particle radius,
one based on CLS and the other on a study of Brownian motion by
DLS. The two results differ by about 4%, rather greater than the

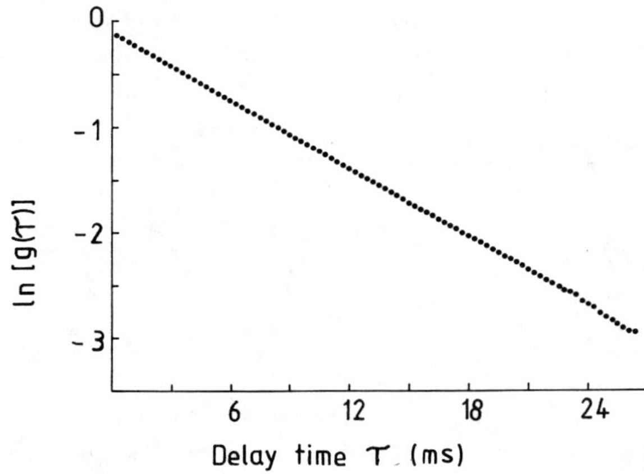

Figure 2

estimated combined error in the measurements. Possible reasons
for this are discussed in reference 21. In any case 4% represents
a fairly typical uncertainty in this type of measurement. It
should also be recognized that, in general, the methods measure
different radii. Thus, in analyzing the Mie data, we have assumed
the particle to be homogeneous. If this were not the case, a complicated average over the distribution of matter within the particle would be obtained. DLS measures the effective radius R_H of
the "hydrodynamic unit". A real particle can be expected to carry
about one or two monolayers of liquid so that R_H could be some $\overset{o}{A}$
greater than the radius of the "bare" particle. This could be a
significant effect for small particles but unimportant for the
large particles considered here.

5   Concluding remarks

A brief summary has been given of the principles and uses of
light scattering methods in colloid science. Properly executed,
light scattering is potentially a clean, accurate and non-pertur-
bative technique. DLS is already established as an important
method for particle sizing; increasing use can be expected in the
study of more complicated particles, slow dynamic processes (e.g.
particle growth and aggregation[22]), interacting particles, electrophoresis etc.

Experimental difficulties in light scattering include: (i) the presence of "dust", unwanted scattering material in the sample; (ii) multiple scattering (the analysis of this article has assumed throughout that a photon is scattered once, at most, on passage through the sample); (iii) absorption of light and subsequent heating of and convection in the sample; (iv) sedimentation. For larger particles, whose Brownian motion is small, the extra particle motions caused by both (iii) and (iv) can be problematical. The effects of and ways to resolve these difficulties are discussed in some of the literature cited in section 1 (see, also, reference 21).

We conclude with a brief comparison between light scattering and neutron scattering. The use in colloid science of neutron scattering, whose underlying principles and theory are very similar to those of light scattering, is growing rapidly with the availability of dedicated facilities such as those in Grenoble (see the article on Neutron Scattering in this volume). The most obvious difference between neutron and light scattering is the wavelength of the radiation, $\lambda \approx 1$ to $10\ \overset{\circ}{A}$ for neutrons, $\lambda \approx 5000\ \overset{\circ}{A}$ for light. This tends to limit neutron scattering to the study of smaller structures than light scattering. However, the optimum size R in a scattering experiment is determined roughly by the criterion $QR \approx 1$ (equation 6), and, by use of very small angle (therefore small Q) instruments, it is possible to study by neutron scattering structures well into the colloidal size range. A major advantage of neutron scattering is that the scattering powers of materials can be varied over wide ranges by isotopic substitutions. Such "contrast variation" is much harder to achieve in light scattering. Because of these two factors elastic neutron scattering is becoming a very versatile probe of structure (time-averaged properties).

However, because there is no neutron equivalent of the laser and because typical motions in colloidal systems are slow compared with those usually studied by inelastic neutron scattering, DLS still provides by far the more precise and detailed data on dynamic properties.

Thus it is best to view neutron scattering and light scattering as complementary rather than competing techniques. Recent studies[23] of the structure of a microemulsion by neutron scattering and of the dynamics of the same system by DLS show the value of this approach.

References

[1] M. Kerker, "The Scattering of Light", Academic Press, New York, 1969.

[2] C. Tanford, "Physical Chemistry of Macromolecules", John Wiley, New York, 1961.

[3] D. McIntyre and F. Gormic, Eds., "Light Scattering from Dilute Polymer Solutions", Gordon and Breach, New York, 1964.

[4] N.A. Clark, J.H. Lunacek and G.B. Benedek, Amer. J. Phys., 1970, 38, 575.

[5] N.C. Ford, Chemica Scripta, 1972, 2, 193.

[6] B. Chu, "Laser Light Scattering", Academic Press, New York, 1974.

[7] H.Z. Cummins and E.R. Pike, Eds. "Photon Correlation and Light Beating Spectroscopy", Plenum, New York, 1974.

[8] B.J. Berne and R. Pecora, "Dynamic Light Scattering", John Wiley New York, 1976.

[9] P.N. Pusey and J.M. Vaughan in "Dielectric and Related Molecular Processes - Volume 2", Ed. M. Davies, The Chemical Society, London, 1975.

[10] H.Z. Cummins and E.R. Pike, Eds., "Photon Correlation Spectroscopy and Velocimetry", Plenum, New York, 1977.

[11] J.M. Schurr, CRC Critical Reviews in Biochemistry, 1977, 371.

[12] B. Chu, Physica Scripta, 1979, 19, 458.

[13] K.J. Randle, Chemistry and Industry, 1980, p 74.

[14] G.N. Ramachandran, Proc. Ind. Acad. Sci. A, 1943, 18, 190.

[15] E. Jakeman, P.N. Pusey and J.M. Vaughan, Optics Communications, 1976, 17, 305.

[16] S.W. Provencher, J. Hendrix, L. De Maeyer and N. Paulussen, J.Chem.Phys., 1978, 69, 4273; N. Ostrowski, D. Sornette, P. Parker and E.R. Pike, Optica Acta, 1981, 28, 1059.

[17] For a review see P.N. Pusey and R.J.A. Tough, in "Dynamic Light Scattering and Velocimetry", Ed. R. Pecora, Plenum, New York, in press.

[18] For a review see B. Ware, Adv. Coll. Interface Sci., 1974, 4, 1.

[19] For a review see H.Z. Cummins in Ref. 10, p 200.

[20] For a review see H.L. Swinney and D.L. Henry, Phys.Rev.A, 1973, 8, 2586.

[21] G.L. Paul and P.N. Pusey, *J.Phys.A:Math.,Gen.*, 1981, <u>14</u>, 3301.

[22] A.R. Goodall, K.J. Randle and M.C. Wilkinson, *J.Coll.Int.Sci.*, 1980, <u>75</u>, 493.

[23] D.J. Cebula, R.H. Ottewill, J. Ralston and P.N. Pusey, *J.Chem.Soc., Faraday Trans. 1.*, 1981, <u>77</u>, 2585.

7
Small Angle Neutron Scattering

By R. H. Ottewill

DEPARTMENT OF PHYSICAL CHEMISTRY, SCHOOL OF CHEMISTRY, UNIVERSITY OF BRISTOL, CANTOCK'S CLOSE, BRISTOL, BS8 1TS, U.K.

Introduction

The examination of colloidal dispersions by some form of radiation has many advantages in that in the volume illuminated the scattered radiation is collected from ca. 10^7 particles, even for a very dilute dispersion. Hence, a large statistical sample is examined. Moreover, the energy imparted to the systems is usually small so that samples can often be illuminated for long periods of time without fear of chemical damage. In addition, the availability of a wide range of detectors, radiation wavelengths and angles of detection makes scattering techniques very versatile.

Probably the most utilised form of radiation is that of visible light. It is widely available in the wavelength range, 400 to 650 nm, using, for example, a mercury arc plus filters, or in highly monochromatic form, various types of laser. The electromagnetic nature of light polarises the outer electrons of the atoms of which the particle is composed and this electronic polarisability is directly related to the refractive index of the material. Consequently, light scattering is dependent on the ratio of the refractive index of the particle to that of the dispersion medium and the size and shape of the particle. The technique is widely used for particle sizing, molecular weight determination etc.,[1] and has the advantage that the basic equipment is readily available and transportable.

There are, however, a number of instances where a shorter wavelength is required and also where different properties of the atoms need to be exploited to obtain information about various parts of the particle and to study, for example, molecules

adsorbed on to particles. It is in these aspects that small
angle neutron scattering offers many advantages over the more
conventional light scattering. In this article, a short account
will be given of the application of small angle neutron scattering
to the study of colloidal dispersions. For more detailed accounts
the reader is referred to standard texts on the subject[2,3,4,5,6].

The Basis of Small Angle Neutron Scattering

The wavelength of a neutron beam is given by the de Broglie
relationship,

$$\lambda \;=\; h/mv$$

where h = Planck's constant, m the mass of the neutron and v = the
velocity of the neutron. Since h and m are fixed, control of the
wavelength can be exercised by moderating the velocity of the
neutrons, and hence the higher the velocity of the neutron, the
shorter the wavelength. In a high flux nuclear reactor such as
that at the Institut Laue Langevin, Grenoble[7], a range of neutron
velocities can be obtained, from thermal neutrons giving $\lambda \sim 1.0$ Å
to cold neutrons giving the wavelength range 4 to 20 Å.

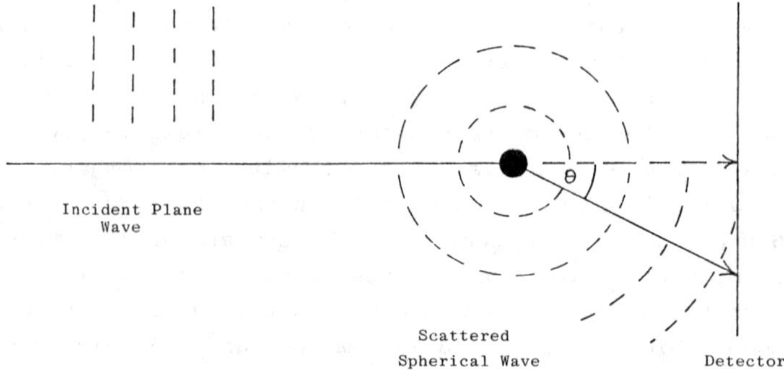

Figure 1:   Schematic illustration of scattering of a neutron
            beam by a single atom

Scattering from a single atom can be represented schematically as shown in Figure 1. Essentially, a planar wave, represented by $\psi(z) = \exp(i\,\underline{k}_o\,z)$, is incident upon the particle and is
then rescattered by the particle in all directions in the form of
a spherical wave, which can be represented by $\psi(\underline{r}) = \dfrac{b_o}{r}\cdot\exp(i\,\underline{k}\,\underline{r})$.
$\underline{k}_o$, the wave vector of the neutron, is related to its wavelength
by

$$|\underline{k}_o| \;=\; \frac{2\pi}{\lambda}$$

The scattered energy is distributed over the surface of a sphere of surface area $4\pi r^2$, the minimum value of which is $4\pi b_o^2$. The quantity b_o is known as the neutron scattering length and since neutrons are scattered by the nucleus of the atoms b_o is of the order of 10^{-12} cm for all nuclei.

The scattering cross-section, σ_s which is the outgoing flux of neutrons divided by the incident flux is given by

$$\sigma_s = 4\pi b_o^2$$

The quantity σ_s is composed of two contributions arising from coherent and incoherent scattering. The latter contribution arises as a consequence of neutron spin and nuclear spin and will not be considered in this article; it does not exist for zero-spin nuclei, e.g. ^{12}C and ^{16}O, but is very important for ^{1}H. This article will be devoted to a treatment of coherent scattering where the coherent scattering cross-section is given by

$$\sigma_{coh} = 4\pi b_{coh}^2$$

In dealing with particles the coherent scattering is dependent on the spatial distribution of the atoms and will depend on the quantity

$$\sum_{i,j} b_i b_j \, \exp\left[i\underline{Q} \cdot (\underline{r}_i - \underline{r}_j)\right]$$

the sum being extended over all pairs of atoms with coherent scattering lengths $b_{i,j}$ with positions defined by $r_{i,j}$. The quantity Q is the scattering vector defined by

$$\underline{Q} = \underline{k} - \underline{k}_o$$

the magnitude of which is given by

$$|Q| = \frac{4\pi}{\lambda} \sin(\theta/2).$$

In what follows it will simply be denoted by Q.

<u>Measurement of Neutron Scattering</u>

Although in principle it would be possible to measure the intensity of radiation scattered at many points over the surface of a sphere, in practice it is more usual to use a planar or near planar detector and to measure the intensity of radiation scattered as a function of the angle θ to the direction of propagation of the incident radiation. The basic experimental arrangement is shown schematically in Figure 2. In practice the range of θ values available on a small angle neutron scattering instrument is between a few minutes and ca. 17°. The variability of the sample

to detector distance, on some machines from a metre or so to 40 m,
allows variation of the angular range and the angular resolution.
The use of multidetectors allows the neutrons arriving at the
various detector cells to be counted and stored in arrays on a
computer disc or tape for subsequent analysis.

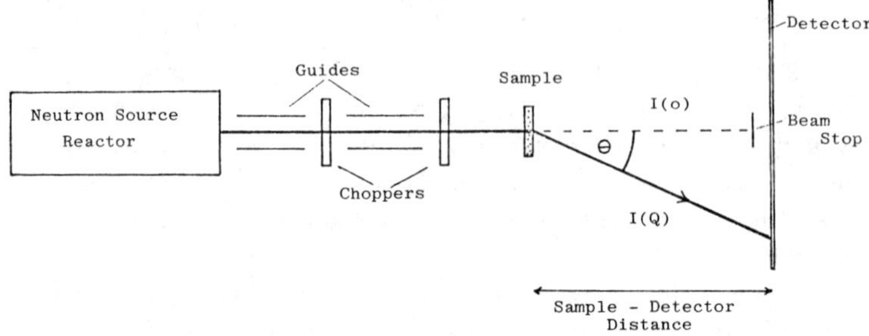

Figure 2:   Schematic diagram of small angle neutron scattering
             apparatus

The intensity of scattering per unit solid angle, frequently
normalised by the scattering from water, is expressed as a function
of the scattering vector Q.   In this article the intensity of
scattering at a particular value of Q will be denoted by I(Q).
Provided that the frequency of the scattered radiation is the
same as that of the incident radiation the scattering is termed
elastic and only this type of scattering will be considered here.

Neutron Scattering Lengths

The term b_{coh} has already been defined as the scattering
length of a particular atom.   We now return to this quantity and
give in Table 1 the values of the coherent scattering length of
a number of atoms[3] taking b_{coh} = b.

A number of points can be noticed from Table 1.   Firstly,
the scattering lengths do not scale with atomic number.   Hence,
D has a very different scattering length from H;   this is a
factor which can be exploited experimentally to considerable
advantage. Secondly, the b value for H is negative, a consequence
of the phase shift of π between waves scattered by H and by D.
Moreover, the scattering from H atoms is predominantly incoherent
and H has a large incoherent scattering length;   for discussion of
this point the reader is referred to standard texts[3,4,5,6].
Thirdly, the neutron scattering length does not depend on
scattering angle.

Table 1

Coherent Scattering Lengths for Some Atoms

Atom	Nucleus	Coherent Scattering Length $b/10^{-12}$ cm
Hydrogen, H	^{1}H	$-$ 0.374
Deuterium, D	^{2}H	0.667
Carbon, C.	^{12}C	0.665
Oxygen, O	^{16}O	0.580
Fluorine, F	^{19}F	0.560
Sulphur, S	^{32}S	0.280

Scattering Length Density

In the field of Colloid Science the particles which act as the scattering units will be composed of an assembly of molecules. For example, in the case of a polystyrene latex particle, the molecular units will be those of styrene, represented by

$$\left[\quad \text{CH} - \text{CH}_2 \quad \right]_n$$

A scattering length density can hence be defined by integrating over all the scattering lengths in the molecular volume, i.e.

$$\rho(r) = \frac{1}{V} \int^{V} b(r).dV$$

or more simply for a single unit of a polystyrene chain the integration can be replaced by a summation to give

$$\rho_{ps} = \frac{\sum b_i}{V}$$

whence taking 8 x 0.665 x 10^{-12} cm for the 8 carbon atoms and 8 x -0.3742 x 10^{-12} cm for the 8 hydrogen atoms and dividing by the molecular volume of 1.64 x 10^{-22} cm^3 we obtain

$$\rho_{ps} = 1.42 \times 10^{10} \text{ cm}^{-2}$$

The molecular weight of the unit was taken as 104.1 and the density of polystyrene[8] as 1.054 g cm^{-3}.

Neutron scattering length densities can be similarly calculated for other molecules and in Table 2 a collection of values is given for the systems discussed in this article.

Table 2

Neutron Scattering Length Densities of a Number of Molecules

Molecules	$\rho/10^{10}\ cm^{-2}$
Water, H_2O	-0.56
Heavy Water, D_2O	6.40
h-polystyrene	1.41
d-polystyrene	6.47
h_{14}-hexane	-0.58
h_{18}-octane	-0.53
d_{18}-octane	6.43
h-polymethylmethacrylate	1.07
d-polymethylmethacrylate	7.03
h_{35}-stearic acid	-0.06
d_{23}-dodecanoic acid	5.30

Scattering from Spherical Colloidal Particles

For a single homogeneous spherical particle, after summing over $b_{i,j}$ and $r_{i,j}$, the intensity of neutrons scattered at a particular value of Q can be expressed by

$$I(Q) = A\ (\rho_p - \rho_m)^2\ Vp^2.\ P(Q) \qquad \ldots \quad (1)$$

where A = an instrument constant, ρ_p = the neutron scattering length density of the particle, ρ_m = the neutron scattering length density of the medium and Vp the volume of the particle as given by

$$Vp\ =\ \frac{4}{3}\ \pi R^3 \qquad \ldots \quad (2)$$

with R = the particle radius. P(Q) is the particle form factor, which for spheres of radius R is given by

$$P(Q) = \left[3(\sin QR - QR\ .\ \cos QR)/(QR)^3 \right]^2 \qquad \ldots \quad (3)$$

Provided that there is no interaction between the particles[9] then for a number concentration of Np particles for unit volume we can write

$$I(Q) = A.9(\rho_p - \rho_m)^2\ Np.Vp^2 \left[\frac{\sin QR - QR.\cos QR}{Q^3R^3} \right]^2 \qquad \ldots \quad (4)$$

Moreover, the volume fraction of a dispersion, ϕ, is given by

$$\phi\ =\ Np\ Vp \qquad \ldots \quad (5)$$

so that equation (4) can be rewritten as

$$I(Q) = A.9(\rho_p - \rho_m)^2 \; \phi \, Vp \left[\frac{\sin QR - QR \cdot \cos QR}{Q^3 R^3}\right]^2 \quad \ldots \quad (6)$$

The form of the curves for completely monodisperse spherical
particles with R = 25 nm and 100 nm are shown in Figure 3.

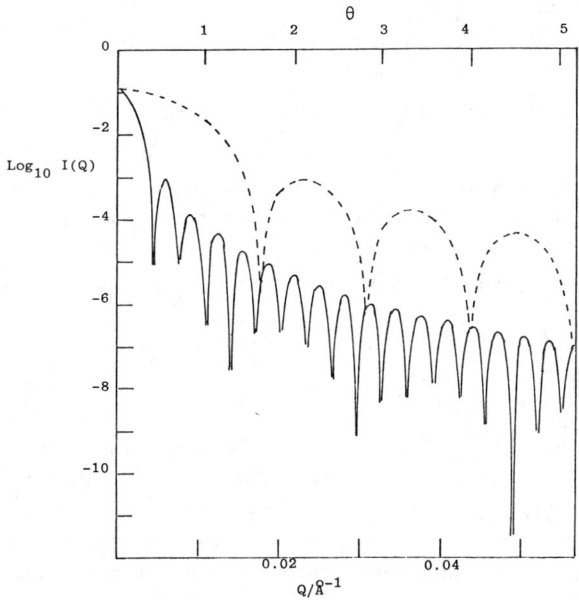

Figure 3:　　Log$_{10}$|I(Q)| against Q for spherical particles
　　　　　　calculated using equation (6). - - - -, R = 25 nm;
　　　　　　————— , R = 100 nm

As can be seen, a number of maxima and minima in intensity occur
in the scattering curves and these increase in number as the
particle size increases. With increase in particle size the first
minimum moves closer to the ordinate and for large particles is
so close to the axis that observations become experimentally
inaccessible. This is quite an advantage since dust particles,
the bane of light scattering, in general do not interfere
seriously with neutron scattering results. The results shown in
Figure 3 are for a completely monodisperse system and resolution
of such high quality is difficult to achieve in practice owing to
particle size polydispersity and other effects (see later). It

should also be noted that the ordinate is plotted as $\log_{10}$ (intensity) so that for larger particles the fall off in intensity between the first and second peaks is several orders of magnitude. The abscissa at the top of the figure is scaled in terms of the scattering angle θ to emphasize the small angular range over which the scattering features need to be observed.

<u>Scattering at Zero Angle</u>

At zero scattering angle, i.e. $Q = 0$, the particle form factor $P(Q)$ becomes unity and hence,

$$I(0) = A (\rho_p - \rho_m)^2 \phi \, Vp \qquad \ldots \quad (7)$$

An immediate use of this relationship is to determine the neutron scattering length of particles. The neutron scattering length density of the medium, ρ_m, can be varied from that of pure water, -0.56×10^{10} cm^{-2}, to that of pure D_2O, 6.40×10^{10} cm^{-2}, by the use of $D_2O:H_2O$ mixtures. Rewriting the equation in the form

$$\pm\sqrt{I(0)} = (\rho_p - \rho_m) \left[\phi \, A \, Vp \right]^{\frac{1}{2}} \qquad \ldots \quad (8)$$

we see immediately that for

$$\rho_p = \rho_m$$

then $\sqrt{I(0)} = 0$. Hence, extrapolating $I(Q)$ against Q data to zero Q enables a plot to be obtained of $\sqrt{I(0)}$ against ρ_m of the form shown in Figure 4 for experiments carried out on h-polystyrene and d-polystyrene. The figures obtained by experiment are in good agreement with those obtained by calculation and listed in Table 2.

From equation (8) we find also that the slope of the curve is given by

$$\frac{d\sqrt{I(0)}}{d\,\rho_m} = - \left[\phi \, A \, Vp \right]^{\frac{1}{2}} \qquad \ldots \quad (9)$$

and the intercept by

$$\rho_p \left[\phi \, A \, Vp \right]^{\frac{1}{2}}$$

whence the ratio of the intercept to the slope is given by

$$\frac{\text{Intercept}}{\text{Slope}} = \rho_p \qquad \ldots \quad (10)$$

Moreover, if the instrument constant is known and the volume fraction of the dispersion then $Vp^{\frac{1}{2}}$ can be obtained from the slope and hence the radius of the particles deduced.

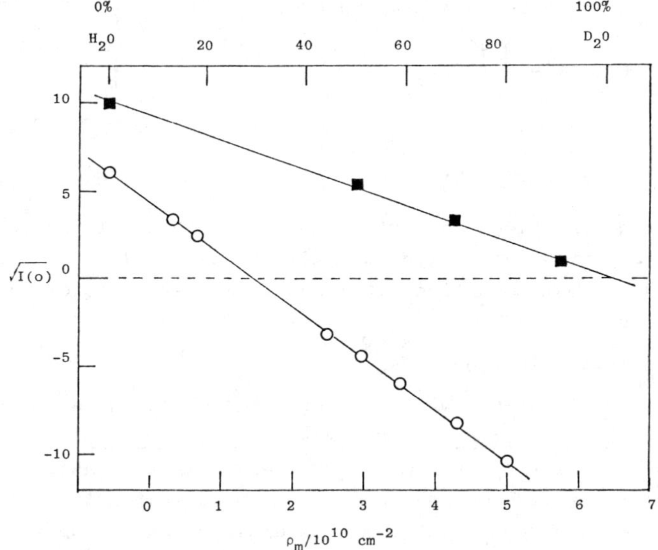

Figure 4: $\sqrt{I(0)}$ against scattering length density of dispersion medium, ρ_m

—O—, polystyrene latex; —■—, polydeuterostyrene

Theory for Small QR Values

For small values of QR the terms in P(Q) can be expanded to give from equation (4)

$$I(Q) = A \; \phi \; Vp(\rho_p - \rho_m)^2 \left(1 - \frac{Q^2R^2}{10} \; \ldots \ldots \right)^2 \qquad \ldots \quad (11)$$

However, since, from the last section,

$$I(0) = A(\rho_p - \rho_m)^2 \; Vp$$

equation (11) can be rewritten as

$$I(Q) = I(0) \left(1 - \frac{Q^2R^2}{10} \; \ldots \ldots \right)^2 \qquad \ldots \quad (12)$$

or in terms of the radius of gyration R_g, since for a sphere,

$$R_g^2 = \frac{3}{5} R^2 \qquad \ldots \quad (13)$$

we obtain

$$I(Q) = I(0) \left(1 - \frac{Q^2 . R_g^2}{2.3} \; \ldots \ldots \right)^2 \qquad \ldots \quad (14)$$

whence expanding by the binomial theorem we obtain

$$I(Q) = I(\theta)\left(1 - \frac{Q^2 R_g^2}{3} + \frac{1}{2^2}\left(\frac{Q^2 R_g^2}{3}\right)^2 \cdots\right) \cdots \quad (15)$$

thus giving

$$I(Q) = I(0)\exp\left(-\frac{Q^2 R_g^2}{3}\right) \qquad \cdots \quad (16)$$

an equation often called the Guinier Law[2,6]. It should be noted,
however, that it is only valid at small values of QR (<< 1). When
applied under these conditions it leads to linear plots of ln I(Q)
against Q^2, viz:

$$\ln I(Q) = \ln I(0) - \frac{R_g^2 . Q^2}{3} \qquad \cdots \quad (17)$$

from the slope of which a value can be obtained for R_g and hence R.

Experimental Investigations of Polystyrene Latex Particles

Polystyrene latices provide very useful dispersions for many
fundamental investigations since they can be prepared in monodisperse form, i.e. essentially as dispersions containing spherical particles with a very narrow size distribution. Moreover,
the particles are amorphous, hence essentially isotropic, and can
be characterized by a single scattering length density.

The experimental results obtained on three latices, containing
particles of diameters 35 nm, 132 nm and 202 nm are shown in
Figures 5, 6 and 7. These data show that with the smallest
particle size a clear first peak is obtained at Q = 0 and a rather
flat second peak over the Q range examined (0.004 to 0.036 $\mathring{A}^{-1}$).
In the case of the 132 nm particles four distinct peaks are
observed but as anticipated from the theoretical plot of Figure 3
a lower Q range (0.0013 to 0.012 $\mathring{A}^{-1}$) has to be utilised to observe
them. With the 202 nm diameter particles (Figure 7) three peaks
are clearly visible and the commencement of the fourth can just be
seen. It can also be estimated that the difference in intensity
between the zero angle peak and the minimum after the third peak
is of the order of four orders of magnitude. It is clear that as
a qualitative means of particle sizing the number of maxima in a
particular Q range can be used for a rapid estimation.

It becomes evident from the experimental results shown in
Figures 5, 6 and 7 that the deep minimum shown in the theoretical
calculations of Figure 3 are not observable in the experimental
curves. There are several reasons for this and they can be
categorised as follows:-

a)   polydispersity of the particle radius,

b)   polychromaticity of the incident neutron beam,

c)   the finite size of the detector elements, i.e. 1 cm x 1 cm,

d)   the angular divergence of the incident beam.

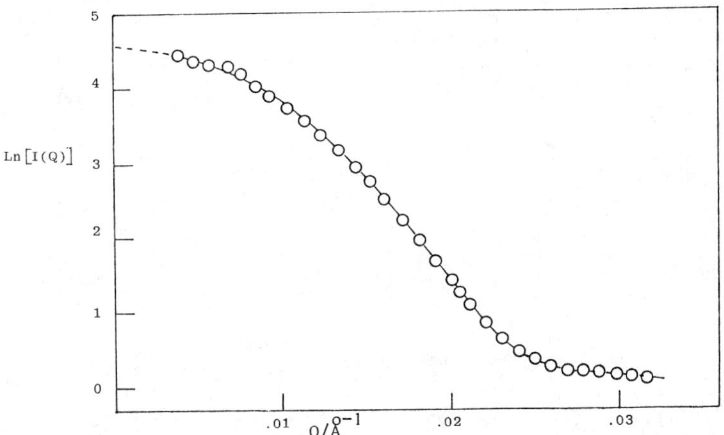

Figure 5:   Ln I(Q) against Q for a polystyrene latex in H_2O.
$\bigcirc$, experimental points; ——— , fitted curve for
$2\bar{R}$ = 34.6 nm and σ = 16%; – – – , fitted extrapolation to
I(0).   Sample-detector distance = 5.53 m; λ = 11.84 Å

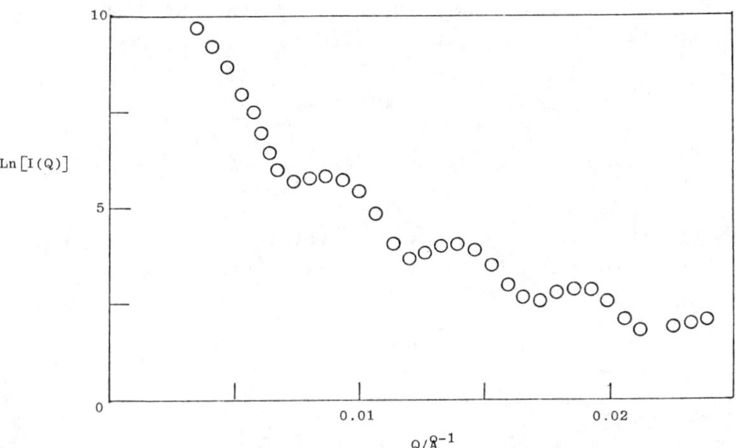

Figure 6:   Ln I(Q) against Q for a polystyrene latex in 70% D_2O:
30% H_2O by volume. $2\bar{R}$ = 132 nm; σ = 4%.
Sample-detector distance = 10.53 m; λ = 9.0 Å

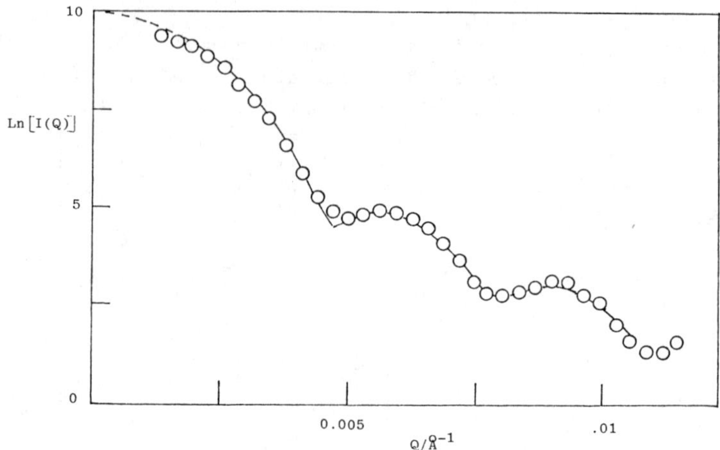

Figure 7: Ln I(Q) against Q for a polystyrene latex in 90% D_2O:
10% H_2O by volume. O , experimental points; ——— ,
fitted curve for $2\bar{R} = 204$ nm and $\sigma = 2.5\%$; – – – ,
fitted extrapolation to I(0). Sample-detector
distance = 40.53 m; $\lambda = 4.98$ Å

The first of these effects is well known in light scattering
and numerous experiments have shown that the particle size
distributions of polymer latices and many other systems can be
well-fitted by a zeroth order logarithmic distribution of the
form[1],

$$p(R) = \frac{\exp\left[-\dfrac{(\ln R - \ln R_m)^2}{2\sigma_o^2}\right]}{(2\pi)^{\frac{1}{2}}\,\sigma_o\,R_m\,\exp(\sigma_o^2/2)} \qquad \ldots \quad (18)$$

where R_m = the modal radius of the particles and σ_o = a parameter
which gives a measure of the width and skewness of the distribution. The relationship between $\bar{R}$, the mean value of R and R_m, is
given by

$$\ln \bar{R} = \ln R_m + 1.5\,\sigma_o^2 \qquad \ldots \quad (19)$$

and the standard deviation σ by,

$$= R_m\left[\exp(4\,\sigma_o^2) - \exp(3\,\sigma_o^2)\right]^{\frac{1}{2}} \qquad \ldots \quad (20)$$

This leads to

$$\sigma = R_m \sigma_o \left[1 + \frac{7 \sigma_o^2}{2!} + \frac{37 \sigma_o^4}{3!} + \ldots \right]^{\frac{1}{2}} \qquad \ldots \quad (21)$$

whence for $\sigma_o \ll 1$

$$\sigma \approx R_m \sigma_o$$

It follows therefore that for a narrow distribution ($\sigma_o \ll 1$) σ_o approximates to the coefficient of variation, viz:

$$\sigma_o \approx C = \frac{\sigma}{\bar{R}} \approx \frac{\sigma}{R_m}$$

The distribution given by equation (18) is particularly suitable for computer processing of data.

The polychromaticity of the neutron beam and the finite size of the detector elements are dependent on the particular instrument used. The data presented in Figures 5, 6 and 7 were obtained on diffractometer D11 at the Institut Laue Langevin, Grenoble. For this instrument with normal collimation (Brunhilde) the wavelength distribution is approximately triangular with a half-width at half-maximum, $\delta\lambda/\lambda$, of 9%[10]. The resolution in terms of Q at the detector can be represented by a Gaussian function with a standard deviation dependent on the sample-detector distance and the wavelength distribution. These instrumental factors have been discussed in detail by Ibel[10], Chauvin[11] and Harris[12].

Once the instrumental factors have been ascertained the computer fitting procedures can be devised using the zeroth order logarithmic function, equation (18), to match the experimental data. Comparisons between the experimental data and computer fits are shown in Figures 5 and 7. The results so obtained can be compared with electron microscope results as shown in Table 3. The agreement is reasonable but the small angle neutron scattering consistently gives slightly smaller radii than electron microscopy.

Table 3

Particle Size Determination

Small Angle Neutron Scattering		Electron Microscope	
$\bar{R}$/nm	σ/%	$\bar{R}$/nm	σ/%
17.3	16		
66.0	4	73.5	8.0
102.0	2.5	103.5	7.9

Particles with an Attached Layer

The case of a spherical particle of radius R_1 and a neutron scattering length density of ρ_p can now be extended to that of the same particle with a concentric layer of neutron scattering length density, ρ_A, such that the overall diameter of the particle becomes R_2. This situation is illustrated in Figure 8.

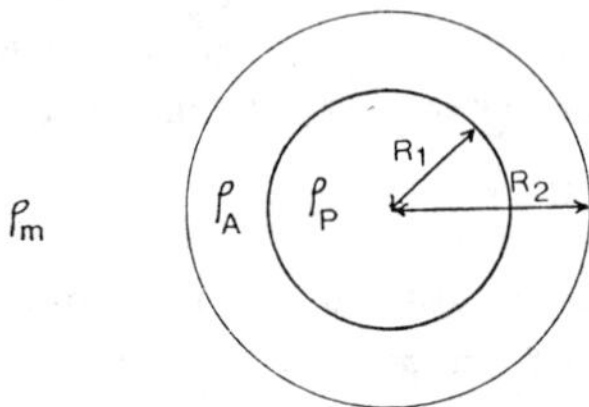

Figure 8:   Spherical particle with a neutron scattering length density of ρ_p and radius R_1, with an attached layer of neutron scattering length density. ρ_A, overall radius R_2, in a medium of neutron scattering length density, ρ_m.

The intensity of scattering for this situation is given by,

$$
\begin{aligned}
I(Q) \;=\; A.Np \, \frac{16\pi^2}{9} \Bigg[\, (\rho_A - \rho_m) \Bigg\{ 3R_2^3 \left(\frac{\sin QR_2 - QR_2 . \cos QR_2}{Q^3 R_2^3} \right) \\
- 3R_1^3 \left(\frac{\sin QR_1 - QR_1 . \cos QR_1}{Q^3 R_1^3} \right) \Bigg\} \\
+ (\rho_p - \rho_m) \, 3R_1^3 \left(\frac{\sin QR_1 - QR_1 . \cos QR_1}{Q^3 R_1^3} \right) \Bigg]^2 \quad \ldots \quad (22)
\end{aligned}
$$

It follows from this equation that if the scattering length density of the medium is made equivalent to that of the attached layer, i.e. $\rho_A = \rho_m$, we obtain

$$
I(Q) = A.Np \, \frac{16\pi^2}{9} \left[(\rho_p - \rho_m) \, 3R_1^3 \left(\frac{\sin QR_1 - QR_1 . \cos QR_1}{Q^3 R_1^3} \right) \right]^2 \quad \ldots \quad (23)
$$

which is identical with equation (6). Under these conditions, therefore, the size of the core particle can be obtained, that is, R_1. However, if ρ_p is known, and it frequently is, then by selecting a suitable dispersion medium of deutero and hydro-compounds it can be arranged that $\rho_m = \rho_p$, thus giving

$$I(Q) = A.Np \frac{16\pi^2}{9} \left[(\rho_A - \rho_m) \left\{ 3R_2^3 \left(\frac{\sin QR_2 - QR_2 . \cos QR_2}{Q^3 R_2^3} \right) - 3R_1^3 \left(\frac{\sin QR_1 - QR_1 . \cos QR_1}{Q^3 R_1^3} \right) \right\} \right]^2 \quad \dots \quad (24)$$

and hence the scattering is that of the attached layer, that is
the system scatters as a shell of scattering length density ρ_A
and thickness $R_2 - R_1$. The possible scattering situations are
illustrated in Figure 9.

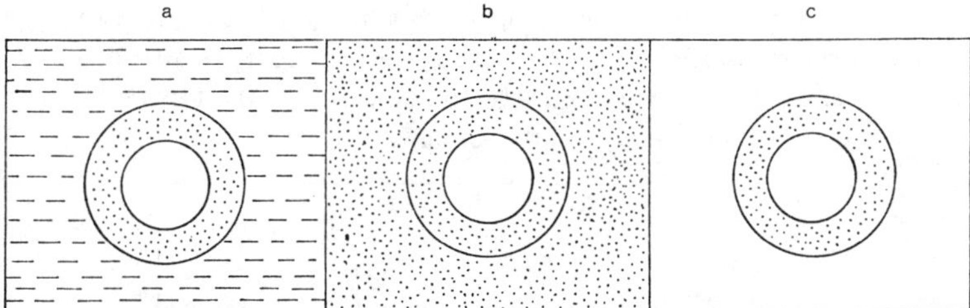

Figure 9: Schematic illustration of contrast matching conditions

a) $\rho_m \neq \rho_A \neq \rho_p$
b) $\rho_m = \rho_A$
c) $\rho_m = \rho_p$

An example of this approach can be provided from data
obtained on a poly-deuteromethyl methacrylate particle stabilised
by a chemically grafted layer of poly-12-hydroxystearic acid[13,14].
The latter layer provides the steric stabilisation necessary to
maintain the particles as discrete entities in a hydrocarbon
environment. The form of the particle is illustrated in Figure 10.

Figure 11 shows curves of I(Q) against Q obtained in h_{14}-
hexane, in order to contrast match the poly-12-hydroxystearic
acid. A good fit to the experimental data was obtained using an
R_1 value of 18 nm[11]. Another series of experiments was carried
out using d_{18}-octane as the dispersion medium, a close match for
the scattering length density of the poly-deuteromethyl methacrylate core of the particle, thus giving the scattering from
a concentric shell in the manner described by equation (24).
The scattering data in this case were fitted with R_1 = 18 nm and
R_2 = 24 nm, thus giving the apparent thickness of the adsorbed

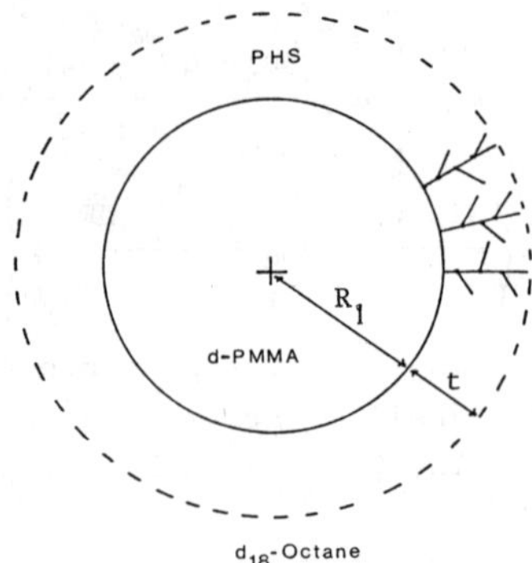

Figure 10:   Schematic representation of a poly-deuteromethyl
             methacrylate (d-PMMA) latex particle of radius R_1,
             stabilised by poly-hydroxystearic acid (PHS) chains
             of thickness t, in d_{18}-octane.

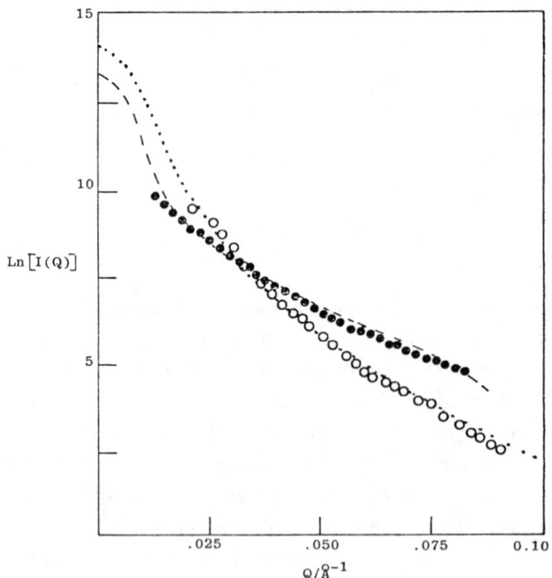

Figure 11:   S.A.N.S. results obtained:- ● , in d_{18}-octane;
             ○ , in h_{14}-hexane.- - - , calculated curve for
             d-PMMA cores of R = 18 nm with t ≈ 6 nm in d_{18}-octane;
             •••• , calculated curve for d-PMMA cores R = 18 nm,
             t = 0 in h_{14}-hexane.

layer, $R_2 - R_1$, as 6 nm. A fully extended poly-12-hydroxystearic acid chain is considered to have a length of ca. 9 nm[13], so that the estimated thickness of the shell from small angle neutron scattering is slightly less than this. On the other hand[9] interaction studies with this system showed the hard sphere radius to be 28 nm, suggesting an adsorbed layer thickness of 10 nm. The difference is almost certainly a consequence of the penetration of solvent into the attached layer of poly-12-hydroxystearic acid.

Additional information on the properties of attached layers can also be obtained by making use of the data obtained at $Q = 0$. For this condition equation (22) becomes

$$I(0) = A.Np \frac{16\pi^2}{9} \left\{ (\rho_A - \rho_m)(R_2^3 - R_1^3) + (\rho_p - \rho_m) R_1^3 \right\}^2 \quad \ldots (25)$$

whence

$$\pm\sqrt{I(0)} = \frac{4\pi}{3}\left[A\,Np\right]^{\frac{1}{2}} \left\{ (\rho_A - \rho_m)(R_2^3 - R_1^3) + (\rho_p - \rho_m) R_1^3 \right\} \quad \ldots (26)$$

If a series of experiments are carried out in mixed hydrocarbon and deuterocarbon dispersion media the position of contrast match can be obtained. At this point $I(0) = 0$ and hence

$$\left[\rho_m\right]_{I(0)=0} = \rho_m^o = \frac{\rho_A R_2^3 - (\rho_A - \rho_p) R_1^3}{R_2^3} \quad \ldots (27)$$

Alternatively, if ρ_m^o, ρ_p, R_1 and R_2 are known, then ρ_A can be determined from the equation in the form

$$\rho_A = \frac{\rho_m^o R_2^3 - \rho_p \cdot R_1^3}{R_2^3 - R_1^3} \quad \ldots (28)$$

As another means of expression if we put V_A = the volume of the attached layer, V_T = the total volume of the particle, i.e. core plus attached layer, and Vp = the volume of the core particle, it follows that

$$\rho_m^o = \frac{\rho_A \cdot V_A + \rho_p \cdot V_p}{V_T} \quad \ldots (29)$$

From a plot of $\sqrt{I(0)}$ against ρ_m it also follows that the slope is given by

$$\frac{d\sqrt{I(0)}}{d\,\rho_m} = -\frac{4}{3}\pi R_2^3 \left[ANp\right]^{\frac{1}{2}} = -V_T\left[ANp\right]^{\frac{1}{2}} \quad \ldots (30)$$

and the intercept at $\rho_m = 0$ by

$$\frac{4}{3}\pi\left[ANp\right]^{\frac{1}{2}}\left[\rho_A(R_2^3 - R_1^3) + \rho_1 R_1^3\right]$$

or in terms of volumes by

$$\left[ANp\right]^{\frac{1}{2}}\left[\rho_A \cdot V_A + \rho_p \cdot V_p\right]$$

which gives the ratio

$$\frac{\text{Intercept}}{\text{Slope}} = \frac{\rho_A \cdot V_A + \rho_p \cdot V_p}{V_T} \qquad \dots \quad (31)$$

In the case cited above of a poly-deutero methylmethacrylate particle with a layer of poly-12-hydroxystearic acid attached by chemical grafting it is not possible to examine the core particle alone. However, in other cases it is possible to start with a bare particle and then to adsorb molecules on to the particle. A case in point is the adsorption of d_{23}-dodecanoic acid onto polystyrene latex particles which has been investigated in some detail by Harris[12,16]. Initially the latex particles (R_1 = 21.3 nm) were examined in H_2O-D_2O mixtures as the dispersion medium and then in the same mixtures with an adsorbed layer of d_{23}-dodecanoic acid. The results are given in Figure 12.

It will be noticed from Figure 9 that there is a cross-over point for the two curves. At this point $I(0)$ is the same for both systems, so that from equations (7) and (25) we obtain

$$(\rho_A - \rho_m)(R_2^3 - R_1^3) + (\rho_p - \rho_m) R_1^3 = (\rho_p - \rho_m) R_1^3 \quad \dots \quad (32)$$

and consequently

$$\rho_A = \left[\rho_m\right] \text{ cross-over}$$

From Figure 12 the information given in Table 4 can readily be obtained.

Table 4

Material	Intercept on ρ_m axis $(\rho_m)_o/\text{cm}^{-2}$	Slope $\dfrac{d\sqrt{I(0)}}{d\rho_m}$	Ratio Intercept Slope/cm^{-2}	ρ_m/cm^{-2} at Cross-Over Point
Polystyrene Latex alone	1.42×10^{10}	-11.96	1.43×10^{10}	4.25×10^{10}
Polystyrene Latex plus d_{23}-dodecanoic acid	1.85×10^{10}	-14.38	1.85×10^{10}	

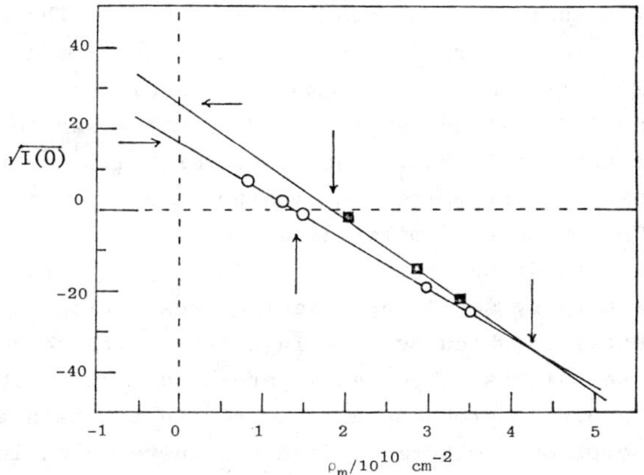

Figure 12: $\sqrt{I(0)}$ against ρ_m for determination of adsorption of dodecanoic acid. —O— , polystyrene latex particles; —■— , polystyrene particles with an adsorbed layer of d_{23}-dodecanoic acid

The radius of the bare latex particles was determined from analysis of the I(Q) against Q curves as 21.3 nm, whence from the ratio of the slopes it was found that the radius of the particle plus an adsorbed layer of d_{23}-dodecanoic acid was 22.7 ± 0.5 nm. Using this information to calculate Vp, V_T and $V_A = V_T - Vp$, and using ρ_p and ρ_m^o a value of 4.38×10^{10} cm^{-2} is obtained for ρ_A; this compares favourably with the value of 4.25×10^{10} cm^{-2} obtained at the cross-over point of the curves.

The value of ρ_A thus obtained is, however, a composite one since the adsorbed layer will contain some solvent. Hence, we can write

$$\rho_A = \rho_{DA} \cdot \alpha + \rho_m^o (1 - \alpha)$$

and since ρ_{DA} = the neutron scattering length for d_{23}-dodecanoic acid (5.3×10^{10} cm^{-2}) and ρ_A and ρ_m^o are known, we obtain α = 0.73 and the volume of the adsorbed layer occupied by d_{23}-dodecanoic acid as 5.08×10^6 Å^3. Moreover, since the molecular volume of d_{23}-dodecanoic acid is 459 Å^3 and the surface area per particle is 5.70×10^5 Å^2, we can readily calculate that there are 1.1×10^4 d_{23}-dodecanoic acid molecules adsorbed per particle and that the area occupied per molecule is 51 ± 5 $\overset{o}{A}{}^2$. These preliminary experiments were carried out at a sodium dodecanoate concentration of 2 x 10^{-2} mol dm^{-3} in 10^{-2} mol dm^{-3} sodium chloride solution at pH 9.2. It was anticipated from the earlier data of Connor[17] that this would be in the monolayer region of the isotherm; the area per molecule compares favourably with that obtained by Connor under similar conditions.

These results demonstrate the feasibility of obtaining adsorption results by small angle neutron scattering and that results are obtained which compare favourably with those obtained by classical techniques. The latter are readily available in most laboratories and will probably always provide the main way of obtaining adsorption isotherms. In cases where molecules are chemically grafted to the surface, however, or the system becomes unstable colloidally in the absence of the adsorbed molecule, then it is clear that small angle neutron scattering provides an excellent means of obtaining information on the thickness of adsorbed layers and the concentration of molecules in them.

Conclusion

A brief review has been given of the application of small angle neutron scattering to a number of problems in the general area of Colloid Science. Another important application is in the study of concentrated dispersions. Consideration of this topic is given later in this volume[9].

<u>References</u>

1. M. Kerker, "The Scattering of Light and other Electromagnetic Radiation", Academic Press, New York, 1969.

2. A. Guinier and G. Fournet, "Small Angle Scattering of X-rays", John Wiley, New York, 1955.

3. G.E. Bacon, "Neutron Scattering in Chemistry", Butterworths, London, 1977.

4. W. Marshall and S.W. Lovesey, "Theory of Thermal Neutron Scattering", Clarendon Press, Oxford, 1971.

5. B.J.M. Willis, "Chemical Applications of Neutron Scattering", Oxford University Press, Oxford, 1973.

6. B. Jacrot, <u>Rep.Prog.Phys.</u>, 1976, <u>39</u>, 911.

7. "Neutron Research Facilities at the High Flux Reactor of the ILL", Institut Max von Laue - Paul Langevin, Grenoble, France, 1975.

8. J.B. Bateman, E.J. Weneck and O.C. Eshler, <u>J.Colloid Science</u>, 1959, <u>14</u>, 308.

9.   R.H. Ottewill, this volume, Chapter 9.

10. K. Ibel, <u>J.App.Crystallography</u>, 1976, <u>9</u>, 296.

11. C. Chauvin, "Thèse de Doctorat", Université de Grenoble, France, 1979.

12. N.M. Harris, D.Phil. Thesis, Oxford University, 1980.

13. R.J.R. Cairns, R.H. Ottewill, D.W.J. Osmond and I. Wagstaff, <u>J.Colloid and Interface Science</u>, 1976, <u>54</u>, 45.

14. K.E.J. Barrett, "Dispersion Polymerization in Organic Media", John Wiley and Sons, London, 1975.

15. D.J. Cebula, J.W. Goodwin and R.H. Ottewill, in press.

16. D.J. Cebula, R.K. Thomas, N.M. Harris, J. Tabony and J.W. White, <u>Faraday Discuss.Chem.Soc.</u>, 1978, <u>65</u>, 76.

17. P. Connor, Ph.D. Thesis, University of Bristol, 1967.

8

Some Uses of Rheology in Colloid Science

By J. W. Goodwin

DEPARTMENT OF PHYSICAL CHEMISTRY, SCHOOL OF CHEMISTRY, UNIVERSITY OF BRISTOL, CANTOCK'S CLOSE, BRISTOL BS8 1TS, U.K.

Introduction

In many industries the colloidal state is chosen for the products or for the production process. Naturally occurring raw materials may also be obtained as colloids. Although this finely divided form may be convenient for achieving the required chemical reactivity, it is often the heat and mass transfer properties of the system that dictate the use of colloidal materials. As a result, the deformation and flow or rheological properties of dispersions are widely studied. A large amount of work is done with quality control aims in mind and this type of work must of necessity be rapid and straight forward to carry out. Usually, only one instrument is used and the behaviour of the material over a much wider range of conditions is taken to be predictable from the information on hand. However problems often arise when extensive changes in formulations are made which alter the inter-action between particles in the dispersions. How the rheological properties change with changes in the type and magnitude of particle interactions is therefore of great importance. The majority of the following discussion will be restricted to solid particles dispersed in a liquid.

Most of the experimental work on suspensions reported in the literature is based either on the measurement of shear stress, τ_{12}, as a function of the rate of deformation or shear rate, $\dot{\gamma}$ (where Newton's dot signifies the time derivative). Some typical rheograms of the types found with colloidal dispersions are shown in Figure 1a. There may also be a time dependence of shear stress at a constant shear rate and this is shown in Figure 1b. During a continuous shearing process, viscoelastic

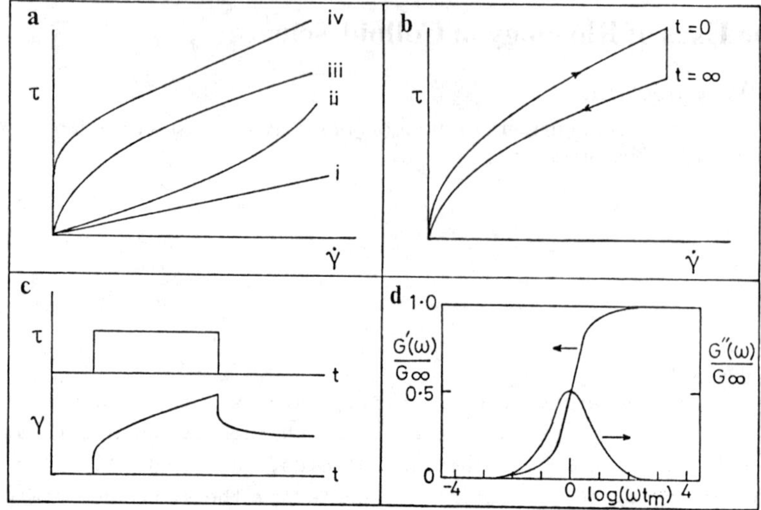

Figure 1.   Typical Rheological Behaviour of Colloidal
                 Dispersions.

    a.   (i) Newtonian;   (ii) shear thickening;
       (iii) pseudoplastic;   (iv) plastic

    b.   thixotropic

    c.   creep response

    d.   dynamic properties;   relative storage and loss
       moduli as a function of dimensionless frequency;
       t_m is the Maxwell relaxation time.

materials should produce normal stresses i.e. τ_{11}, τ_{22} and τ_{33}
and although the measurement of the normal force (the first
normal stress difference, $(\tau_{11} - \tau_{22})$) has been frequently studied
in polymer rheology it has been largely neglected with colloidal
dispersions. Viscoelastic dispersions have usually been
characterised by utilising oscillatory measurements to give the
complex modulus:

$$G^* = G' + i\,G'' \qquad\qquad \dots\quad (1)$$

where G' is the storage modulus and G" is the loss modulus. The
frequency dependence of these yields information on the relaxation
processes that are occurring. Figure 1d illustrates a dimensionless

plot of the storage and loss moduli as a function of frequency.
G_∞ is the high frequency limit of G' and t_m is the Maxwell
relaxation time.

It is not intended that this should be an extensive review of
dispersion rheology but rather a guide to some of the more
important problems to which recent work has provided some solutions, at least in part. The concept of the structure in a
dispersion, that is the spatial distribution of particles, and
the modification of the structure due to interparticle forces is
central to all the work considered here. Typical curves for
charge stabilised particles are shown in Figure 2 for the pair
potential, V_T, and the force between two particles, F_T, as a
function of the surface-to-surface separation, h. A detailed

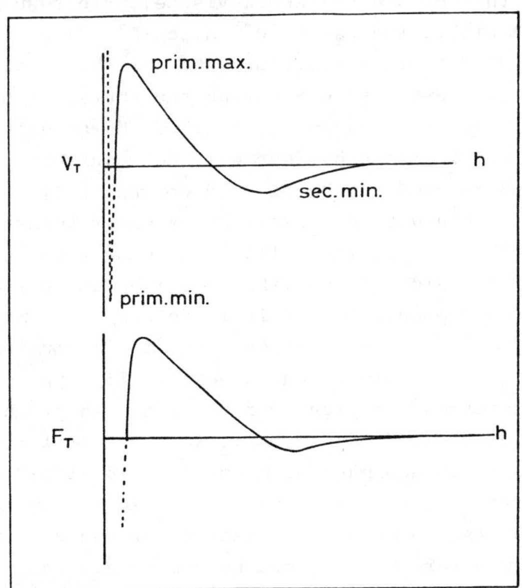

Figure 2:   Potential energy and force as a function of
              interparticle separation for electrostatically
              stabilised particles.

discussion of both these forces and steric forces is to be found
elsewhere in this volume. It is hoped that this chapter will
indicate some of the ways that rheological measurements can be
used to characterise both the forces and the particles in
colloidal dispersions.

Experimental Methods

Perhaps the most important consideration in the choice of an experimental method is the time scale of the laboratory experiment compared to that of the process that one is attempting to describe. The concept of the Deborah number[5], D_e, makes the reason for this requirement quite clear. If the experimental time is denoted by t_e and the stress relaxation time, or characteristic time of the material, by t_r then

$$D_e = \frac{t_r}{t_e} \qquad \qquad \ldots \quad (1)$$

Elastic solids are therefore materials with $D_e > 1$ and viscous liquids have $D_e < 1$. Now many colloidal systems lie between these limits so that $D_e \sim 0\ (1)$ and a viscoelastic response is obtained. Times within the range 10^{-4} s to 10^4 s are conveniently handled in the laboratory and, if the relaxation processes are sufficiently slow to bring the characteristic time of the materials into this range, a detailed characterisation of the resulting viscoelastic behaviour will be required. The experimental methods used can be divided conveniently into continuous shear techniques and oscillatory shear techniques.

Continuous Shear: Capillary flow can be used to measure viscosity. As a technique it has the following advantages. It has a high degree of precision and it is both a cheap method and a simple one to operate. In the U-tube type of viscometers, the applied stress is gravitational but an externally applied pressure, such as compressed air, may also be used. The "Gun Rheometer" (Deimos Limited, Canterbury) is a recent introduction of this type and it is designed to measure yield stress by a progressive increase in applied pressure until the sample in a capillary tube begins to move. The rate of pressure increase must be specified before results can be compared. The disadvantages of capillary flow are that there is a wide range of shear rates across a capillary tube and that the time dependence of thixotropic materials cannot easily be measured.

Rotational viscometers such as cone and plate or concentric cylinder instruments overcome these last two disadvantages of capillary instruments, so that for a constant angular rotation, shear stress can be measured as a function of both shear rate and time. (Instruments in which a small diameter cylindrical bob is immersed in a large reservoir, and which can be taken as the case of unbounded flow, allow the time dependence only to be followed). There are a large number of instruments currently available which fulfil this role admirably and the choice of instrument will normally be dependent on what other facilities are required; for example, whether or not the first normal stress difference that occurs when viscoelastic material is sheared needs to be measured.

The measurement of the strain or strain rate as a function of applied stress can be conveniently measured with a "Deer Rheometer" (Rheometer Marketing Limited, Leeds) in which the stress is applied directly via a servomotor with the central shaft mounted in air-bearing. Stresses as low as 0.1 N m^{-2} can be applied to a dispersion and so the apparatus is particularly suitable for creep compliance measurements. A schematic creep curve is shown in Figure 1c. This is typical of the behaviour of most viscoelastic dispersions.

Oscillatory Shear: A parallel plate geometry is perhaps the simplest platen arrangement but a cone and plate or concentric cylinders may also be used. In this type of experiment a small sinusoidal strain is applied continuously and the resultant stress is recorded. The phase shift and amplitude ratio are then used to give the complex modulus which is the sum of storage (elastic) and loss (viscous) moduli. The Weissenberg Rheogoniometer, the Instron 3250 Rheometer and the Rheometrics 4 system may all be operated in this mode in addition to continuous shear. The latter two instruments have load cell arrangements for the measurement of stress which do not have a resonant frequency within the applied frequency window as can occur with the Rheogoniometer, although this instrument has been in use for a considerable period and has proved its worth with polymer solutions and melts.

Polymer solutions give a linear viscoelastic response with much larger strains than particulate dispersions. The rate of change of interparticle force with distance (the curvature of the pair potential) is markedly non-linear and so very small strains must be applied if a linear response is to be found. Measurement of the propagation velocity of a shear wave can also be used to study the viscoelastic properties of some dispersions. The "Shearometer" (Rank Brothers, Bottisham, Cambridge) is designed to do this. A very small strain is applied by means of a piezo-

electric crystal and the frequency of the transmitted wave is in
the region of 200 Hz. This measurement gives the shear wave
rigidity modulus, $\tilde{G}$, which is related to the storage modulus, G',
of the dispersion. For many weak gels the characeristic time is
much longer than the period of the wave, i.e. t_r >> 0.005 s, and
there is little attenuation of the wave. Under these conditions
the wave rigidity modulus is equal to the high frequency limit
of the storage modulus so that:

$$\tilde{G} = \underset{Lt\ \omega\ \to\ \infty}{G'} = G_\infty$$

Extensional Flow: Measurement of the extensional viscosity
of polymer solutions has recently been receiving considerable
attention and a commercial instrument is now available (Sangamo-
Schlumberger, Bognor Regis). For Newtonian materials, Trouton
showed that the value of the extensional viscosity was three
times that of the shear viscosity. However, for viscoelastic
materials the value of the extensional viscosity can become
several orders of magnitude higher than that of the shear value
and so, in principle, an important insight into the mechanism
producing viscoelasticity might be gained by this measurement.
Experimentally the method available relies on having a spinnable
material and so only dispersions in polymer solutions can be
investigated at present.

Dispersions of Hard Spheres

With dispersions we wish to know how the hydrodynamic
forces that occur during flow are modified by the presence of the
particles. The particle shape and charge, the extent of
aggregation, and the presence of stabiliser layers all have an
effect on the stress and so the measurement of this under
precisely defined flow rates can be used to characterise the
particles in dispersions. Einstein[6] derived the relationship
between the viscosity of a dispersion, η, the volume fraction of
the dispersions, ϕ, and the viscosity of the continuous phase,
η_o, as:

$$\frac{\eta}{\eta_o} = 1 + 2.5\ \phi \qquad\qquad \ldots\ (2)$$

This was derived for dispersions of hard spheres in the limit of
infinite dilution so that all interactions between particles
were neglected. This equation can be used with a negligible
error for ϕ < 0.01 for suspensions of particles which can be

approximated by hard spheres i.e. no <u>long</u> range electrostatic or steric interactions and no aggregation of the spherical particles.

Particle Interactions: In a quiescent colloidal dispersion the particles are moving continuously in a random manner with Brownian motion. An equilibrium statistical distribution is evolved which depends on the volume fraction and interparticle potentials. Using the Einstein – Smoluchowski relation for the time scale of the motion, with the Stokes – Einstein equation for the diffusion coefficient, the time taken for a particle to diffuse a distance equal to its radius is equal to $6\pi \, \eta_o \, a^3/kT$. This characterises the time taken for the restoration of the equilibrium microstructure after a disturbance, caused for example by convective motion, i.e. this is the relaxation time of the microstructure. The time scale of shear flow is given by the reciprocal of the shear rate. The dimensionless group formed by the ratio of these two times:

$$\frac{6\pi \, \eta_o \, a^3 \, \dot{\gamma}}{kT}$$

gives the relative importance of diffusion and convection to the microstructure. For example, if $6\pi \, \eta_o \, a^3 \, \dot{\gamma}/kT \ll 1$ the distribution of particles is only slightly altered by the flow. This case has been studied extensively by Batchelor[7,8,9] who derived the following equation for hard sphere suspensions:

$$\frac{\eta(o)}{\eta_o} = 1 + 2.5\,\phi + 6.2\,\phi^2 + 0\,(\phi^3) \quad \dots \quad (3a)$$

where $\eta(o)$ is the low shear rate limit of the viscosity as Brownian motion is dominant. For extensional flow a similar equation was derived:

$$\frac{\eta(o)}{\eta_o} = 1 + 2.5\,\phi + 7.6\,\phi^2 + 0\,(\phi^3) \quad \dots \quad (3b)$$

At high shear rates, i.e. the dimensionless shear rate ($6\pi \, \eta_o \, a^3 \, \dot{\gamma}/kT \gg 1$), the equilibrium microstructure is significantly altered by the flow, giving rise to the shear thinning behaviour that has been observed experimentally[1,10] with both poly(styrene) and poly(vinyl chloride) latices. These experimental data lie on curves of the form sketched in Figure 3. Equation (3a) was shown to fit Krieger's data for $\phi < 0.15$[4]. It should be noted that the cube of the particle radius appears in the numerator of the expression for the dimensionless shear rate and the rate of observed shear rates over which the viscosity is shear thinning will move to lower

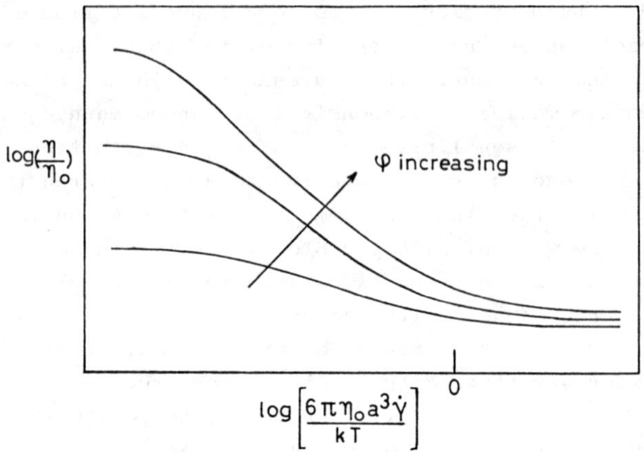

Figure 3:   Relative viscosity of dispersions as a function of
reduced shear-rate.

values as the particle radius increases. Hence there is both a
non-Newtonian behaviour and a particle size dependence on the
viscosity of hard sphere dispersions due to particle interactions
and the microstructure in the dispersions, although this will
result in only a small variation in the value of viscosity at
$\phi < 0.1$ as the terms of $0\,(\phi)^2$ contribute less than 5% to the
viscosity.

As the volume fraction is increased beyond 0.15 a rigorous
hydrodynamic theory is not available as multiparticle interactions become increasingly important, i.e. terms of order ϕ^3 and
ϕ^4 etc. would have to be included. The problem, therefore, must
be treated empirically and there are a great many equations in
the literature to choose from. Mooney[11] predicted an exponential
form from space-filling considerations but Dougherty and Krieger[1]
corrected the derivation and gave the following equation:

$$\frac{\eta}{\eta_o} = [1 - k_p \phi]^{-[\eta]/k_p} \qquad \dots \quad (4)$$

in which $[\eta]$ is the intrinsic viscosity which is 2.5 for hard
spheres (i.e. rigid and uncharged) and k_p is the reciprocal of
the maximum packing fraction at which flow can occur i.e. the
viscosity becomes infinite at $\phi = \dfrac{1}{k_p}$.

It is possible to arrive at this equation very simply using

a mean field argument as was shown by Ball and Richmond[32].
Writing the Einstein equation (Equation 3) with the intrinsic
viscosity, $[\eta]$, and differentiating gives the initial rate of
change of viscosity

$$\frac{\partial \eta}{\partial \phi} = [\eta] \, \eta_o \, .$$

The mean field concept leads us to expect that a similar initial
rate of change of viscosity would occur at any volume fraction
on the addition of a small number of particles, so that:

$$\delta \eta = [\eta] \eta \, \delta \phi$$

However the particles already in suspension occupy a finite
volume and so the last addition of particles is to the free fluid
volume. The excluded volume of any particle is $\frac{4}{3}\pi \, a^3 \, k_p$ and if
there are N_p particles per unit volume of dispersion, the second
addition of particles is to a reduced volume of $1 - N_p \, k_p \, \frac{4\pi}{3} \, a^3$
so that the change in volume fraction is $\frac{\delta \phi}{1 - k_p \phi}$. Hence:

$$\frac{\partial \eta}{\eta} = \frac{[\eta] \, \partial \phi}{1 - k_p \phi}$$

The boundary conditions are

$$\eta \rightarrow \eta_o \text{ when } \phi \rightarrow 0$$

so the solution is

$$\ln \eta = -\frac{[\eta]}{k_p} \ln (1 - k_p \phi) + \ln \eta_o \qquad \cdots \qquad (5)$$

which is Krieger's result in Equation (4). Krieger found
$k_p = f(\dot{\gamma})$ and to vary from 1.47 at the high shear limit to 1.75
at the low shear limit. Ball and Richmond[32] indicated that k_p
might be expected to change with volume fraction but only slowly.
However, anticipating the discussion in the following section, for
both charged and anisometric particles k_p is likely to change
markedly with volume fraction and so in general $k_p = f(\dot{\gamma}, \phi)$.

The other approach to the problem of calculating the viscosity at high concentrations has been to use a cell model. Frankel
and Acrivos[12], Ackerman and Shen[13], and more recently Graham[14],
used a cell model with simple cubic symmetry and calculated the
total energy dissipated as the sum of that dissipated by fluid
flow around a central sphere and that due to interactions between
the spheres calculated by means of lubrication theory. The
resulting expression of Graham's for viscosity as a function of
volume fraction for elongational flow, η_{ex}, with h being the

surface-to-surface separation between spheres of radius a was:

$$\frac{\eta_{ex}(o)}{\eta_o} = 1 + 2.5\ \phi + \frac{9}{4}\ (1 + \frac{h}{2a})\ \frac{1}{(h/a)} - \left[\frac{1}{(h/a) + 1} - \frac{1}{[(h/a) + 1]^2}\right]$$

$$\ldots \quad (6)$$

The relationship given by Ackerman and Shen[13] for shear flow was:

$$\frac{\eta(o)}{\eta_o} = \left(1 - \frac{\pi}{4\beta^2}\right) + \left(\frac{\pi}{4} - \frac{\pi}{6\beta}\right)\left(\frac{1}{\beta^2 - 1}\right)\ \ 1 + \frac{2}{(\beta^2 - 1)v^2}\ \tan^{-1}\left(\left(\frac{\beta+1}{\beta-1}\right)^{\frac{1}{2}}\right)$$

$$\ldots \quad (7)$$

where $\beta = \left(\frac{1}{k_p\phi}\right)^{1/3}$

Equations 4, 6 and 7 all give a good fit to the experimental data compiled by Thomas[15] if the maximum volume fraction at which flow can occur is used as a fitting parameter. Equation (7) is less good than the others at $\phi > 0.5$. Although all of these equations can be faulted on a theoretical basis[9], they can be useful for curve-fitting and it will be interesting to see if the cell models can be modified to give the correct dependence on the colloidal forces acting between particles.

Particles with Adsorbed Layers

Stability of colloidal particles is often produced by the adsorption of molecules on to the surface of the particles. The stabilisers may be ionic surface active agents used to produce electrostatic repulsive forces between particles, non-ionic surface active agents or polymer molecules to provide steric stabilisation or poly-electrolyte molecules to give both electro-static and steric stabilisation. A piece of information that is frequently required in adsorption studies is the extension of the layer normal to surface of the colloidal particles and the measurement of suspension viscosity is sometimes used to calculate this. For a densely packed well defined layer, such as occurs on the formation of a monolayer of an ionic surface active agent, the hydrodynamic volume of each particle is increased by the volume of the adsorbed layer so that the *effective* volume frac-tion ϕ' is:

$$\phi' = \phi(1 + \frac{\delta}{a})^3 \qquad \ldots \quad (8)$$

where δ is the thickness of the layer. Dilute dispersions and capillary viscometry are frequently used for the viscosity measurements and Equation (3) employed after the substitution of Equation (8). Usually $\frac{\delta}{a} \ll 1$ and a small uncertainty in the viscosity results in a large uncertainty in δ, so it can be more useful to carry out measurements over a wide range of volume fractions and use Equation (4) for example. Reference 1 provides a good example of the precision that can be obtained. However it may be quicker and give at least as good precision to use a sedimentation technique or photon correlation spectroscopy to give the hydrodynamic radius of the particle. For extended, low density, polymeric layers the hydrodynamic measurement may not provide a reliable guide to the interparticle separation at which steric forces begin to occur.

<u>Anisometric Particles</u>

If the particles in a suspension are anisometric, the prediction of the intrinsic viscosity, i.e.

$$[\eta] = \left(\frac{\eta/\eta_o - 1}{\phi} \right)_{\phi \to o}$$

which is of course 2.5 for uncharged spheres, is a complicated problem and only for some of the cases are solutions available[7]. The testing of the theories by experiment is also a problem as there are no model systems of monodisperse particles available to date in which both the particle size and axial ratio can be varied systematically. However many systems of great practical importance consist of particles which are markedly anisometric, for example, clay dispersions and blood.

In the absence of Brownian motion, spherical particles in a shear field rotate with the vorticity of that field so that they have a constant angular velocity of $\frac{\dot{\gamma}}{2}$. The angular velocity of ellipsoidal particles varies periodically with the orientation of the particle with respect to the field. The equations of motion for both prolate and oblate spheroids have been derived by Jeffrey[16] assuming the absence of Brownian motion, particle-particle interactions and inertial forces. Figure 4 shows a coordinate system located at the centre of a particle with Φ and θ as the polar angles of the major axis. The rate of change of these angles are :

$$\dot{\Phi} = \frac{\dot{\gamma}}{r_e^2 + 1} (r_e^2 \cos^2 \Phi + \sin^2 \Phi) \qquad \ldots \quad (9)$$

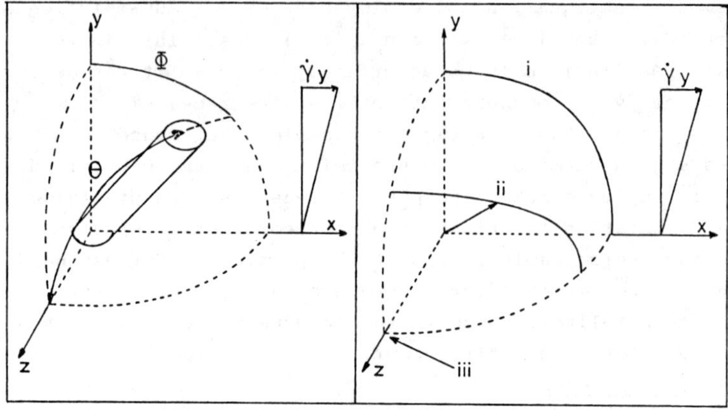

Figure 4:   Coordinate system   Figure 5:   Orbits for a prolate
             for a rotating rod   spheroid in shear
                                                    flow. Orbit
                                                    constants:
                                                    i) C = ∞;
                                                    ii) C = 1;
                                                    iii) C = 0.

$$\tan \Theta = \frac{C\, r_e}{\left(r_e^2 \cos^2 \Phi + \sin^2 \Phi\right)^{\frac{1}{2}}} \qquad \ldots \quad (10)$$

The orbit constant C in Equation (10) defines the ellipse traced
out by the end of the ellipsoid and has a value of 0 when the
a-axis is parallel to the vorticity axis and ∞ when the a-axis
lies in the x-y plane. These orbits are shown in Figure 5.

The period of rotation about the vorticity axis is given by
t* as:

$$t^* = \frac{2\pi}{\dot{\gamma}} \left(r_e + \frac{1}{r_e}\right) \qquad \ldots \quad (11)$$

If t* can be measured, the equivalent ellipsoidal axial ratio
can be calculated for particle shapes other than ellipsoidal, for
example rods whether of circular or rectangular cross-section.

Brownian motion will tend to randomise the angular velocity
and cause drift from one orbit to another. The rotary Péclet
number, Pe_r, gives the relative importance of convective to
Brownian rotation.

$$Pe_r = \frac{\dot{\gamma}}{2\,D_r}$$

where D_r is the rotary diffusion coefficient and $D_r = \dfrac{kT}{f_r a^3 \eta_o}$

where k is the Boltzmann constant, T is the absolute temperature and f_r is the rotary friction coefficient.

For the condition $Pe_r \rightarrow \infty$, Jeffrey[16] calculated the intrinsic viscosity for ellipsoids of low axial ratio and for two particular cases, namely for $C = \infty$ and $C = 0$, corresponding to maximum and minimum energy dissipation. The results are shown in Figure 6.

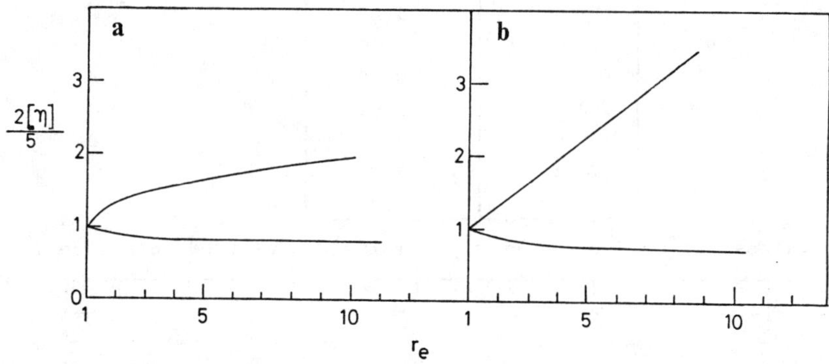

Figure 6:   Effect of shape on the intrinsic viscosity as a function of axial ratio from reference 16.
a) prolate ellipsoids   b) oblate ellipsoids.

Eisenschitz[17] assumed that there should be an isotropic distribution of orbits and for $r_e > 1$ he derived the intrinsic viscosity as:

$$[\eta] \simeq 0.366\,\frac{r_e}{\ln 2r_e} \qquad \ldots \quad (12)$$

With $Pe_r \gg 1$, the effect on particle rotation due to Brownian motion is minimal except that drift from one orbit to another can occur and this ensures an isotropic distribution of orbits. This was the case analysed by Leal and Hinch[18]. It was found that the resulting orbit distribution corresponded to low values of the energy dissipation rate and they gave the following asymptotic expressions for the intrinsic viscosity:

$$\text{for } r_e > 20, \quad [\eta] \simeq \frac{0.312\, r_e}{\ln 2r_e - 1.5} + 2 \qquad \ldots \quad (13)$$

$$\text{and for } r_e < 0.2, \quad [\eta] \simeq 3.183 - 1.792\, r_e \qquad \ldots \quad (14)$$

Table 1 gives their numerical data computed for intermediate axial ratios.

<u>Table 1</u>

Intrinsic viscosity calculated as a function of axial ratio for ellipsoids at $Pe_r \gg 1$

r_e	$[\eta]$
20	4.4
18.4	4.28
10	3.46
3.85	2.78
2.44	2.62
1.05	2.50
0.41	2.60
0.26	2.71

The values of $[\eta]$ for rod-like particles are plotted in Figure 7

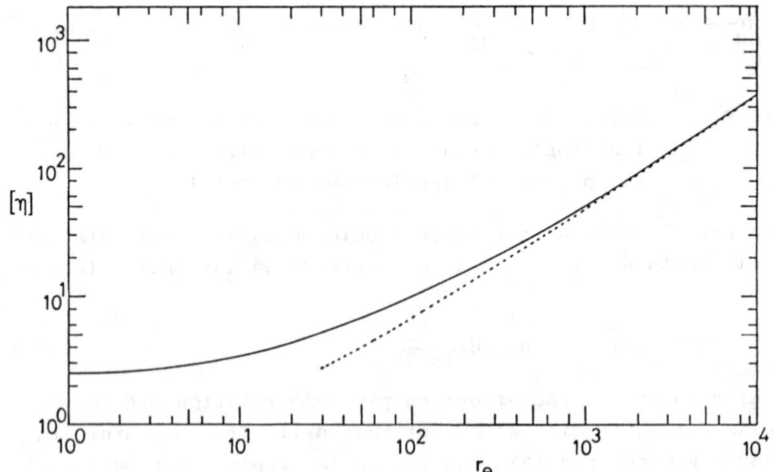

Figure 7:   Intrinsic viscosity as a function of axial ratio for
            prolate ellipsoids
            ———— from Leal and Hinch, reference 18
            .... from Eisenschitz.   Equation (13), reference 17

for values of $r_e < 10^4$. Also shown in this figure are data calculated from Equation (12) which has a somewhat similar form to Equation (13). The analysis by Leal and Hinch appears to be the best available to date for $Pe_r < 1$; however the agreement with experimental data does not appear to be very good in all cases. It has been shown by Gauthier[19] that the effect of fluid inertia is to cause particles to drift into orbits corresponding to maximum energy dissipation i.e. for $C \rightarrow \infty$ for rods and $C \rightarrow 0$ for discs. This became of increasing importance when the particle Reynolds' number, Re_p, which indicates the relative importance of inertial to viscous effects, was in excess of 10^{-2} where:

$$Re_p = \frac{16\ a^2\ \dot{\gamma}}{\eta_o}\ \rho_o$$

where ρ_o is the density of the continuous phase. Hence shear rates in excess of $10^3\ s^{-1}$ will favour high intrinsic viscosities for suspensions of anisometric particles in the colloidal size range. However the effect of collisions between particles has been shown by Anzurowski and Mason[20] to randomise the orbit distribution function and decreasing values of $[\eta]$ were found experimentally[19,20] for both rods and discs as a function of volume fraction. These experimental data are shown in Table 2.

<u>Table 2</u>

Intrinsic viscosities of rods and discs in Newtonian media (from reference 19)

	r_e	Re_p	$\phi \times 10^4$	$[\eta]$	$[\eta]^a$
Rods	15.5	10^{-1}	0.30	4.93	4.0
			0.46	4.72	
			1.0	4.04	
		10^{-3}	0.46	2.52	
Discs	0.139	10^{-1}	13	4.79	2.9
			23	4.03	
			39	3.52	
		10^{-3}	23	2.21	

[a] Calculated from reference 18

Values calculated from the analysis by Leal and Hinch are also shown and it appears that these values might have been approached at high shear rates as the volume fraction was increased. Figure 8

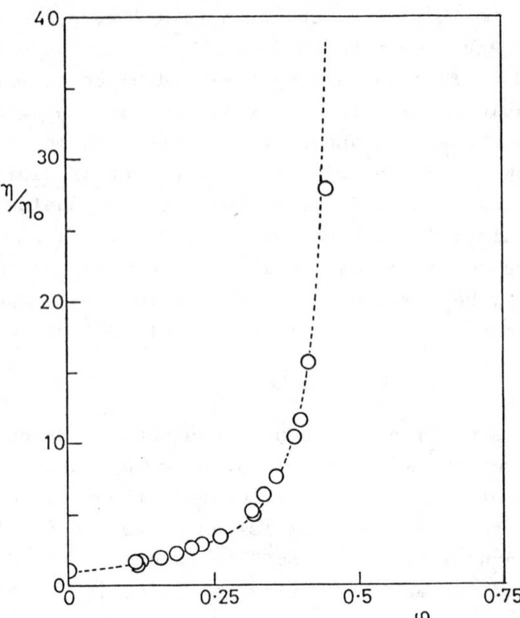

Figure 8:   Relative viscosity as a function of volume fraction
             for hardened red blood cells in 0.145 mol dm^{-3} sodium
             chloride.   0 experimental points from reference 21;
             —— Calculated from Equation (5), $[\eta]$ = 3.16,
             ϕ_m = 0.5.

shows data obtained[21] using human erythrocytes in a 0.145 mol
dm^{-3} sodium chloride solution.   These had been hardened by
acetaldehyde treatment and gave a dispersion of rigid biconcave
discs of r_e = 0.291.   By using Equation (5) to fit the data,
values of k = 2.0 and $[\eta]$ = 3.2 were calculated.   These data were
obtained at high volume fractions ($0.1 < \phi < 0.5$) and the
suspensions all showed Newtonian behaviour at shear rates
corresponding to $Re_p \leq 0.13$.   The value calculated from reference
18 was $[\eta]$ = 2.7 which is significantly lower than the value
found experimentally.   Viscosity data obtained[22] over a volume
fraction range of 0.004 to 0.04 using kaolinite particles
dispersed with a cationic surface active agent corresponded to an
intrinsic viscosity of 3.5 which is appreciably higher than that
calculated for a particle of r_e = 0.12 from reference 18, i.e. $[\eta]$
= 2.9 but similar to the values found at $\phi \approx 0.04$ by Gauthier
et al.[19] of $[\eta]$ = 3.52. However the corresponding value of Re_p

for the kaolinite particles was 10^{-3}.

It is clear from the above discussion that a considerable amount of careful experimental work is still required in this area, and in particular this must take into account the effects of volume fraction and Reynolds' number, before the intrinsic viscosity can be used as a precise characterisation of particle shape. When particles are dispersed in a non-Newtonian medium further complications arise. For example Gauthier et al.[19] have shown that the orbital drift in a system of particles dispersed in a viscoelastic fluid was to orbits corresponding to minimum energy dissipation rates and hence low values of intrinsic viscosity.

Charged Particles

Stable colloidal dispersions in aqueous media frequently owe their stability, at least in part, to the forces of electrostatic repulsion between particles. The origin of the charge may be adsorption of ions, ionic surface active agents or polyelectrolytes, isomorphous substitution in the lattice of crystalline particles, or ionic groups chemically bound to the molecules making up polymeric particles. Three distinct electroviscous effects have been identified[23] and the result of these can be a dramatic change in viscosity with changes in the electrolyte concentration or pH of the suspending medium. This may be seen most markedly with concentrated dispersions. The mechanisms assigned to the three effects are quite different from each other.

The Primary Electroviscous Effect. This is the result of the diffuse part of the electrical double layer that is formed around charged particles being deformed from spherical symmetry by a shear field. The analysis by Booth[24] was carried out for weak flows, i.e. only small distortion of the diffuse double layer and for small electrokinetic potentials, i.e. $\zeta < 25$ mV. The result of the analysis gave the intrinsic viscosity for charged particles in a 1:1 electrolyte as:

$$[\eta] \;=\; \frac{5}{2}\left[1 \,+\, (2\varepsilon\varepsilon_o\zeta)^2\, \frac{(\kappa a)\,(1\,+\,\kappa a)^2\,Z(\kappa a)}{\sigma\,\eta_o\,a^2}\right] \quad \ldots \quad (15)$$

where ε is the relative permittivity, ε_o is the permittivity of free space, σ is specific conductivity of the continuous phase. κ^{-1} is the Debye length which is dependent on the ionic strength, which can be denoted by the symbol I, and then:

$$\frac{1}{\kappa} = \left(\frac{5 \times 10^2 \; \varepsilon\varepsilon_o kT}{N_a \; e^2 \; I} \right)^{\frac{1}{2}}$$

where N_a is the Avogadro number and e is the charge on the electron. The function $Z(\kappa a)$ was a power series with the two limiting
forms, for extended diffuse layers:

$$\text{small } \kappa a; \quad Z(\kappa a) \simeq \frac{1}{200\pi\kappa a} + \frac{11 \; \kappa a}{3200\pi} \qquad \ldots \quad (16a)$$

$$\text{large } \kappa a; \quad Z(\kappa a) \quad \frac{3}{2\pi \; (\kappa a)^4} \qquad \ldots \quad (16b)$$

Substitution of Equation (16b) in Equation (15) gives the Krasny-
Ergen[25] equation which was calculated assuming the diffuse layer
was sufficiently thin for the curvature to be ignored.

Russel[26] also derived the same result as Krasny-Ergen;
however he extended the analysis to stronger flows and made two
important observations. Firstly there should be a shear thinning
contribution and secondly normal stresses should occur. The
ratio of the convective diffusive motion of the ions is given by
the translational Péclet number of the ions Pe_i as:

$$Pe_i = \frac{\dot{\gamma} \; a_i^2}{u_i kT}$$

where a_i is the hydrodynamic radius of the ions and u_i is their
mobility. Clearly the greater the value of Pe_i the greater is
the expected distortion of the diffuse layer. Russel's result
for moderate shear rates was[26]:

$$[\eta] = \frac{5}{2} \left[1 + (\varepsilon\varepsilon_o\zeta)^2 \; \frac{6}{\eta_o \; \sigma \; a^2} \; \frac{1}{(1 + Pe_i^2)} \right] \qquad \ldots \quad (17)$$

and the first normal stress difference was given as:

$$\tau_{11} - \tau_{22} = \frac{30 \; \zeta^2}{a^2} \; \phi \; \frac{Pe_i^2}{1 + Pe_i^2} \qquad \ldots \quad (18a)$$

and the second normal stress difference as:

$$\tau_{33} - \tau_{22} = - \frac{(\tau_{11} - \tau_{22})}{2} \qquad \ldots \quad (18b)$$

Lever[27] considered the case of large distortion of the diffuse
layer i.e. $\kappa a < 1$ and arbitrary Pe_i and worked with both simple
shear and extensional flow. Again viscoelasticity was predicted

and for low ion Péclet numbers the first normal stress difference was dependent on the square of the Péclet number; however at high values of Pe_i, $(\tau_{11} - \tau_{22}) \propto \gamma^{-\frac{1}{2}}$. A numerical solution was found for the shear stress and all the stresses went through a maximum at $Pe_i \sim 10$. Of course with large distortions of the diffuse layer some elastic properties are to be expected. However the primary electroviscous effect is usually small when compared to the other electroviscous effects and it is doubtful if the currently available instrumentation is sufficiently sensitive to measure the predicted normal forces.

<u>The Secondary Electroviscous Effect</u>. The mechanism for this effect has been identified as an increase in the collision diameter of the particles due to electrostatic repulsive forces. As a result, the excluded volume is greater than that for uncharged particles and the electrostatic particle-particle interactions in a flowing dispersion give an additional source of energy dissipation[23,28,29]. Under quiescent conditions the minimum separation between particles and hence the pair distribution function is dependent on the ratio of the net interparticle repulsive forces to those due to Brownian motion. van der Waals' forces will of course be much smaller than the electrostatic forces at surface separations $> \frac{1}{\kappa}$ for particles in dispersions showing marked electroviscous effects and so use of the electrostatic term only provides a good approximation. Russel[29,4] used the following group for the ratio of repulsive to Brownian forces:

$$\alpha = \frac{16\pi^2 \, \varepsilon\varepsilon_o \zeta^2 \, a^4 \, \kappa}{kT} \, \exp(2\,\kappa a)$$

and, for values of $\alpha \gg 1$, gave a collision diameter, L_o, for the particles in the absence of convection forces, that is an equivalent hard sphere diameter, as:

$$L_o = \frac{1}{\kappa} \ln \left[\frac{\alpha}{\ln(\alpha/\ln\,\alpha)} \right]$$

Under shear flow the minimum centre-to-centre separation between particles, L, would be $2a \leq L \leq L_o$, that is at the high shear limit $L = 2a$. Russel's analysis resulted in the following expression for the zero shear rate limit of the relative viscosity:

$$\frac{\eta(o)}{\eta_o} = 1 + [\eta]\,\phi + \frac{2}{5}([\eta]\phi)^2 + \frac{3}{40}\ln\left(\frac{\alpha}{\ln\,\alpha}\right)\left[\ln\left(\frac{\alpha}{\ln(\alpha/\ln\,\alpha)}\right)\right]^4 \frac{\phi^2}{(\kappa a)^5}$$

$$\ldots \quad (19)$$

The intrinsic viscosity includes the primary electroviscous effect. The experimental data of Stone-Masui and Watillon[30] was used to test Equation (19) and a good fit was obtained for the coefficients of the $0(\phi^2)$ term for polymer latices which had values for this coefficient as high as 1.2×10^3.

There are two important consequences of interparticle repulsion on the trajectories of colliding particles. Firstly there are no longer closed trajectories[31,51] so that all doublets will separate and secondly the paths of collision and recession become different from one another. Typical trajectories are shown in Figure 9.

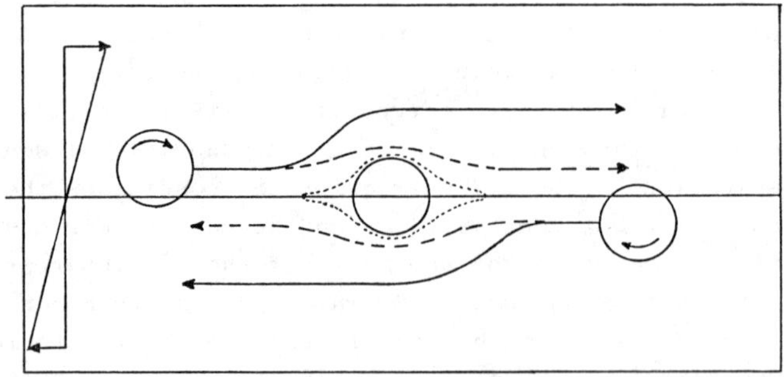

Figure 9:   Orbits for colliding pairs of particles in shear flow
— — —uncharged particles, separating orbits;
········· uncharged particles, closed orbits;
———— charged particles.

<u>The Tertiary Electroviscous Effect</u>. This effect is due to the expansion and contraction of particles made up, at least in part, of polyelectrolytes with changes in the electrolyte concentration or pH of the medium. For dispersions of particles stabilised by a layer of polyelectrolyte, adsorbed or chemically bound to the surface of the particles, the problem resolves to the calculation of the conformation of the stabiliser molecules and the use of Equation (8) for the calculation of the effective volume fraction. The difficulties are in determining the effect of the particle surface on the conformation of the polyelectrolyte chains and the extent of the penetration of moving liquid into the stabiliser layer. Both Evans <u>et al</u>.[33] and Buscall[34] have examined the properties of monodisperse polymer latices stabilised by poly-acrylic acid grafted to surface of the particles and so perhaps

some detailed studies of the tertiary electroviscous effect may
appear in the near future.

<u>Concentrated Suspensions of Charged Particles.</u> It has been
demonstrated with great clarity by Krieger[1,35] that the viscosity
of suspensions of charged particles can be many times that of a
suspension of a similar concentration of hard spheres. The
experiments were carried out as a function of ionic strength
using monodisperse polymer latices. Concentrated latices are
iridescent due to Bragg diffraction in the visible region of the
spectrum and this has been shown[36,37,38,39] to be due to the
formation of a face-centred cubic arrangement of particles.
Brown <u>et al</u>.[40] have shown from light scattering measurements of
the structure factor using relatively dilute but interacting
particles that the radial distribution function of the particles
is similar to that found in liquids. With increasing concentration the degree of order can therefore be expected to increase
from that which is analogous to a gaseous state to an ordered
liquid-like state which exhibits iridescence. The analysis in the
previous section which was due to Russel[29] pertains to the
former of these two states. It has now been established by
several workers[41-46,4] that the ordered liquid-like state is
viscoelastic.

The elastic properties of dispersions of interacting
particles have been measured by three different methods. Compression techniques have been used to measure the bulk modulus of
the array of particles[44,45]. Russel[4] reports measurement of
storage and loss moduli from oscillatory viscometry to give
values of G' and G". Measurement of the propagation velocity of
shear waves[41,42,43] has been used to give the wave rigidity
modulus using moderately concentrated dispersions, i.e. $\phi > 0.15$.
These systems showed little attenuation of the shear wave so that
at the frequency used G' >> G" and the wave rigidity modulus is
equal to the high frequency limit of the storage modulus G_∞
(i.e. $G_\infty = G'_{Lt\ \omega\to\infty}$ where ω is the frequency of the deformation).
In all cases the elastic properties were shown to increase rapidly
with increasing volume fraction.

The problem now becomes how to relate the rheological
properties to the double layer interaction between particles at
the high concentrations. van Megen and co-workers[48] have used
Monte Carlo type simulations to calculate the value of G_∞. Pair-
wise additivity was assumed for the calculation of the interaction

force between any particle and its neighbours and face-centred
cubic symmetry was used to calculate an orientational average
modulus. By treating the dispersion as though the particles
were in a static array, i.e. using a model that can be described
as a "static structure model", it was possible to derive a simple
algebraic expression for the shear modulus[41,42,43] :

$$G_\infty = \frac{0.46\ \phi^{1/3}}{a}\ \frac{\partial^2 V_T}{\partial h^2} \qquad \ldots \quad (20)$$

where the pair potential, V_T, was calculated from the sum of the
electrostatic and van der Waals interaction energies. The numerical value of 0.46 was the result of assuming face-centred
cubic symmetry i.e. using twelve nearest neighbours for the
pairwise estimation of the interaction force and also for
calculating the surface-surface separation, h,viz:

$$h = 2a \left[\left(\frac{\phi_m}{\phi}\right)^{1/3} - 1 \right]$$

where ϕ_m = 0.74. It should be stressed that, although this
static structure model used face-centred cubic symmetry, no long
range order was assumed so that it was compatible with the type
of liquid-like structure that has been observed, albeit analogous
to the state at 0^OK. The model described the experimental data
well[43] and was in quite good agreement with the computer
simulation[48]. Russel[4] reported the results of an analysis using
a self-consistent field approach which appeared to give quite
good agreement with experiment.

The prediction of the viscous behaviour of these concentrated
systems has received less attention. However a model has been
developed recently[49] which predicts the zero shear stress
viscosity and the results were compared with creep compliance
measurements. The model assumed that the total shear stress was
the sum of the hydrodynamic and electrostatic contributions.
Equation (4) was used to give the hydrodynamic contribution, i.e.

$$\tau_H = \dot\gamma\, \eta_o \left[1 - k_p\phi \right]^{-[\eta]/k_p}$$

although Equation (7) could have been as appropriate. The
electrostatic contribution was calculated in the following manner.
Face-centred cubic symmetry was assumed but with a large number
of stacking faults or holes. Although the particles could
oscillate about their mean position, the jump diffusion into a

vacancy was assumed to be slow and governed by the electrostatic interactions. Eyring's[50] analysis using absolute rate theory for simple liquids was used to give the relationship between the shear rate and the electrostatic contribution to the stress, τ_E:

$$\dot{\gamma} = \frac{4\pi kT}{\hbar} \exp(F^*/kT) \sinh(\tau_E r^3/2kT) \qquad \ldots \quad (21)$$

where $\hbar$ is Plancks constant$/2\pi$, r is the centre-to-centre separation between particles and F^* is the activation energy for the jump diffusive motion. At sufficiently low stress Equation (21) was approximated by :

$$\dot{\gamma} = \frac{2\pi}{\hbar} \tau_E r^3 \exp(F^*/kT) \qquad \ldots \quad (22)$$

This gave the total shear stress as:

$$\tau_{12}(O) = \tau_E + \tau_H$$

$$= \frac{\dot{\gamma}\hbar}{2\pi r^3} \exp(-F^*/kT) + \dot{\gamma}\eta_o(1 - k\phi)^{-[\eta]/k_p}$$

and hence:

$$\frac{\eta(o)}{\eta_o} = \frac{\hbar}{2\pi r^3 \eta_o} \exp(-F^*/kT) + (1 - k\phi)^{-[\eta]/k_p} \quad \ldots \quad (23)$$

The activation energy was calculated by assuming pairwise additivity and using the geometry of the array to calculate the energy difference between the position midway between a lattice site and an adjacent vacancy and that of the lattice site itself i.e. $F^* = f(V_T)$. This model was successful in accounting for the very high values of $\eta(o)$ that were found experimentally[49]. It should be stressed that the systems were extremely shear thinning with changes of viscosity by factors of up to 10^7 being observed. This type of analysis is clearly limited to a range of particle and electrolyte concentrations where the pairwise additivity assumption is a reasonable approximation, and also to low stress as it was implicit in the model that particle distribution is not significantly distorted by the applied stress. Figure 10 illustrates the result of calculations using Equation (23).

By carrying out experiments on concentrated latices under conditions where $D_e \gg 1$ (measurement of the wave rigidity modulus) and also $D_e \ll 1$ (measurement of the viscous part of the creep compliance) it is possible to gain an insight into both the interaction energy between particles and the curvature of their energy with respect to particle separation. The problems of

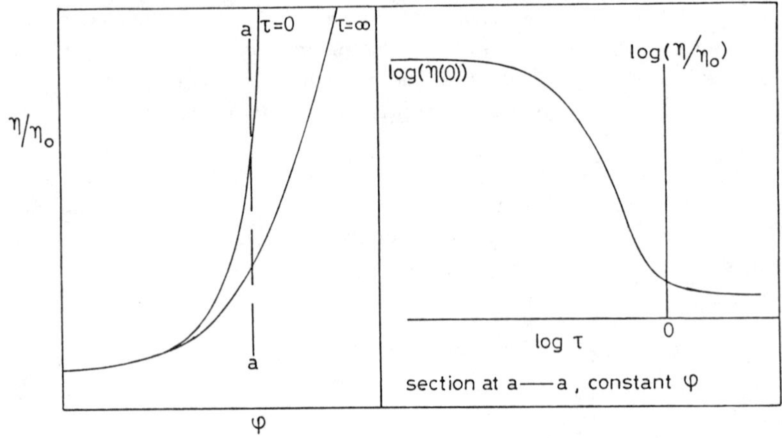

Figure 10:   Relative viscosity as a function of volume fraction
              and applied stress for suspensions of charged
              colloidal particles.

predicting the rheological behaviour of concentrated dispersions
of charged particles at high shear rates, or if the particles
are anisometric or are polydisperse, require a good deal of
further study.

Suspensions of Aggregated Particles

The Effect of Flow Fields on Aggregation.   In the previous
section the effect of strong forces of repulsion was reviewed.
The pair distribution of particles was altered and an increase in
viscosity was predicted.   Attractive forces may dominate the
interparticle forces under some conditions and these will also
have a profound effect on the particle distribution function and
hence on the rheological properties.   A recent advance in the
prediction of the stability of dilute dispersions during flow
has been made by van de Ven and Mason[51-54] and Zeichner and
Schowalter[55,56].   Both groups used a detailed calculation of the
trajectories of colliding spheres to determine the effect of the
flow field on the stability of pairs of particles.   It was
predicted that particles which showed flocculation in a secondary
minimum could be disaggregated by causing the system to flow.   At
high deformation rates orthokinetic aggregation into a primary
minimum was predicted and redispersion could only occur at very
high shear rates.   Some of the data from Reference 55 are plotted

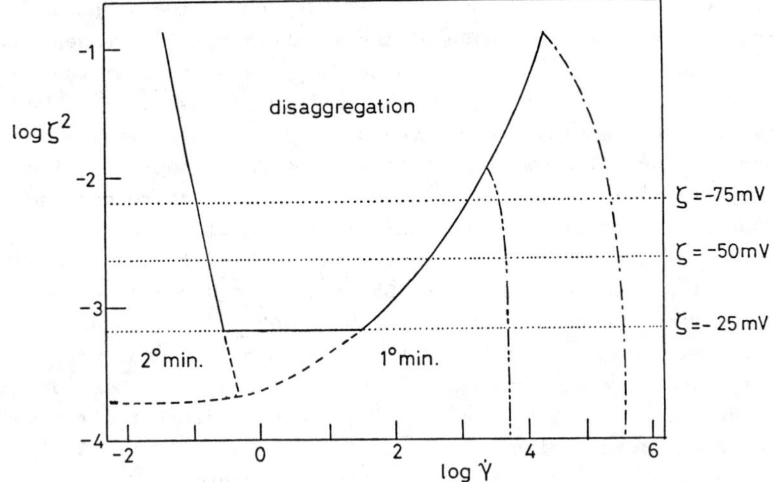

Figure 11:   Shear stability diagram calculated from Ref.55
for a = 1   m, κa = 100, using a retarded force law
with A_{12} = 7 x 10^{-21} J.—··—, h min = 4 nm;
— · —, h min = 0.4 nm.

in a recast form in Figure 11 as an illustration of the predictions that can be made from this type of analysis.   The curves
are applicable to polystyrene latex particles of particle radius
1 μm dispersed in a 10^{-3} mol dm^{-3} sodium chloride solution.
Three distinct regions are shown in which a pair of particles
that collide can:

1) form a stable doublet of particles flocculated in the
    secondary minimum.   This means that there is a finite
    separation between them and so the particles rotate
    independently and this increases the rotational period of the
    doublet compared to two spheres in contact; and

2) at high shear rate, secondary minimum doublets can be
    separated and the system is in disaggregated condition; and

3) when the suspensions are sheared at very high shear rates
    orthokinetic coagulation into a primary minimum state was
    predicted.   The rate of orthokinetic aggregation was
    predicted by both groups to be proportional to $\dot{\gamma}^{0.8}$.
    Disaggregation from this state required much higher shear
    rates due to assymmetry of the interparticle force-distance
    curve.

The curves are of course sensitive to the choice of parameters; for example, the value of the Hamaker function chosen and in particular the shear rate at which disaggregation from the secondary minimum was predicted, was extremely sensitive to the minimum separation distance between particles in a primary minimum. Figure 11 shows two redispersion limits corresponding to a primary minimum location of 0.4 nm and 4 nm, corresponding to adsorbed layers of water or stabiliser molecules such as non-ionic surface active agent, and this resulted in a predicted decrease of a hundred-fold in the shear rate required to cause disaggregation. Shear rates in excess of 10^5 s^{-1} are outside the range of most viscometric systems and so break-up of primary minimum doublets would not be observed easily in the laboratory. With a values of $\zeta < 14$ mV secondary minimum flocculation was not predicted and coagulation was the preferred state. Table 3 summarises the various boundary positions predicted. The values of the shear rates at the boundaries were shown[51,55] to move to higher values at increased electrolyte concentration and with an increase in particle diameter. Reynolds[57] has also confirmed this recently by determining the change in the number of single particles in a dilute latex as a function of time at different shear rates.

<u>Table 3</u>

Values of shear rate producing aggregation-disaggregation of 1 µm diameter latex particles in 10^{-3} mol dm^{-3} sodium chloride predicted from Figure 11.

ζ/mV	Secondary Doublets to Singlets $\dot{\gamma}$/s^{-1}	Singlets to Primary Doublets $\dot{\gamma}$/s^{-1}	Primary Doublets to Singlets $\dot{\gamma}$/s^{-1}	
			h_{min} 4 nm	h_{min} 0.4nm
75	0.1	1,300	5×10^3	2×10^5
50	0.2	330	5×10^3	2×10^5
25	0.3	20	5×10^3	2×10^5

Figure 12 shows his experimental results for shear rates of 30 s^{-1} and 1.5 s^{-1} with 1.8 µm radius polystyrene particles in a 10^{-2} mol dm^{-3} sodium chloride solution. At the higher shear rate some redispersion occurred, whilst at the lowest value aggregation was enhanced.

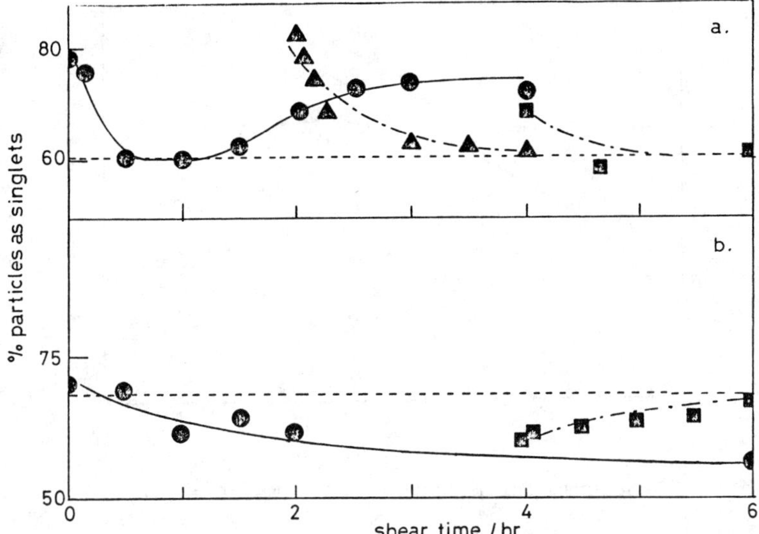

Figure 12:   Extent of secondary minimum aggregation in Couette
Flow for 1.8 μm radius polystyrene. a) Shear rate
30 s^{-1}; b) Shear rate 1.5 s^{-1}; ——— flow;
—·——shear stopped;- - - - - perikinetic control.

The Rheological Properties of Coagulated Dispersions. It is of
considerable importance in many processes to be able to correlate
the viscous behaviour with changes in the chemical environment
which can be expected to alter the coagulation properties.
Plastic or pseudoplastic behaviour is usually observed for these
types of dispersion. A typical curve for shear stress as a
function of shear rate is illustrated in Figure 13 which includes
the definition of the plastic viscosity, η_{pl}, the Bingham yield
stress, τ_B, and the yield stress τ_y. The high shear rate
behaviour has frequently been used as a means of characterising
the structured properties, i.e.

$$\tau_{12} = \eta_{pl} \dot{\gamma} + \tau_B \qquad \qquad ... \quad (24)$$

Michaels and Bolger[58] used this to relate the energy dissipation
in breaking up coagula to the Bingham yield stress. This may be

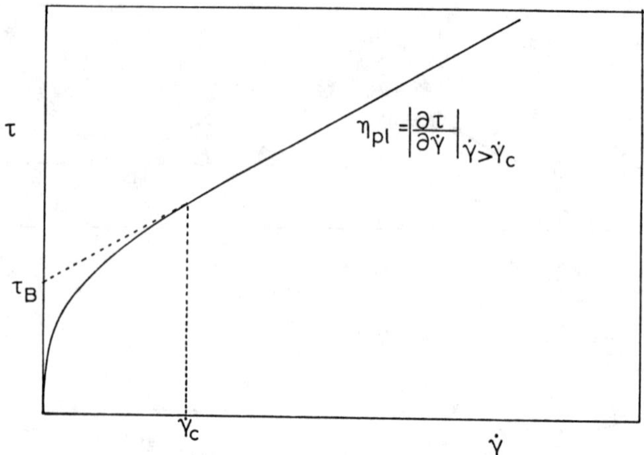

Figure 13:   Rheogram for a pseudoplastic material

achieved by utilising the two-body collision frequency and the interaction energy between the particles calculated at some arbitrary
separation distance considered to be the position of the primary
minimum, h_{min}:

$$\tau_B = \frac{3\alpha_o \, |V(h_{min})| \, \phi_f^2}{\pi^2 \, a_f^3} \qquad \ldots \quad (25)$$

where a_f and ϕ_f refer to the radius and volume fraction of the
flow unit, i.e. the coagulum with its occluded fluid, and α_o is the
orthokinetic capture efficiency which allows for the interparticle
forces and the hydrodynamic interactions reducing the aggregation
rate and has a value of $0 < \alpha_o < 1$. The plastic viscosity was
calculated as the viscosity of a suspension of spheres of volume
fraction ϕ_f and hence ϕ_f may be determined from the data by use,
for example, of Equation (5). If the value of a_f is known then
$V(h_{min})$ can be calculated. From the calculations represented by
Table 3, it is likely that values of $a_f > 1$ μm should occur.
Hunter and co-workers have carried out a considerable body of
work on this problem and perhaps the key publications are with
Firth[54], van de Ven[60] and Frayne[61]. They conclude that by far the
greatest source of energy dissipation is due to the "pumping" of
liquid around inside the coagula as it is continually deformed in

a shear field. The result of their treatment gives:

$$\tau_B = \frac{27\alpha_o \; a_f^2 \; \delta\eta_o \; \dot{\gamma} \; C_{FP}}{5 \; a^3} \; \phi^2 \qquad \qquad \cdots \quad (26)$$

where δ is the increase in interparticle separation during
stretching (~ 0.5 nm) and $C_{FP} = \phi_f/\phi$ and gives the void density
in the coagula ($C_{FP} \sim 10$) and $C_{FP} \propto V(h_{min})$. This model gave
reasonable correlations for τ_B with ζ^2, C_{FP}, η_o, a_f and ϕ^2 using
poly(methyl methacrylate) latices destabilised with a cationic
surface active agent. Perhaps one of the biggest criticisms of
both of these models is that they require a precise estimation
of the pair potential at close distances of separation and this
is the region of greatest uncertainty. However, they can both
be useful in providing correlations between the rheological
behaviour of suspensions and their aggregation state.

1.   I.M. Krieger, <u>Adv.Coll.and Interface Sci.</u>, 1972, <u>3</u>, 111.

2.   J.W. Goodwin, Spec.Periodical Reports, Colloid Sci.2,
         Royal Society of Chemistry, 1975, pp.246.

3.   J. Mewis and A.J.B. Spaull, <u>Adv.Coll.and Interface Sci.</u>,
         1976, <u>6</u>, 173.

4.   W.B. Russel, <u>J.Rheol.</u>, 1980, <u>24</u>, 287.

5.   M. Reiner, <u>Physics Today</u>, 1969, <u>17</u>, 62,

6.   A. Einstein, "Investigation on the Theory of the Brownian
         Movement", Dover, NY. 1956.

7.   G.K. Batchelor, <u>J.Fluid Mech.</u>, 1970, <u>41</u>, 545.

8.   G.K. Batchelor, <u>Ann.Rev.Fluid Mech.</u>, 1974, <u>6</u>, 227.

9.   G.K. Batchelor, <u>J.Fluid Mech.</u>, 1977, <u>83</u>, 97.

10. S.J. Willey and C.W. Macosko, <u>J.Rheol.</u>, 1978, <u>22</u>, 525.

11. M. Mooney, <u>J.Coll.Sci.</u>, 1951, <u>6</u>, 162.

12. N.A. Frankel and A. Acrivos, <u>Chem.Eng.Sci.</u>, 1967, <u>22</u>, 847.

13. N.L. Ackermann and H.T. Shen, <u>A.I.Chem.Eng.J.</u>, 1979, <u>25</u>, 327.

14. A.L. Graham, Rheology Research Centre Report No.62, University
         of Wisconsin, June 1980.

15. D.G. Thomas, <u>J.Coll.Sci.</u>, 1967, <u>20</u>, 267.

16. G.B. Jeffrey, <u>Proc.Roy.Soc.</u>, 1922, <u>A102</u>, 161.

17. R. Eisenschitz, <u>S.Phys.Chemie</u>, 1931, <u>A158</u>, 78.

18. L.G. Leal and E.J. Hinch, <u>J.Fluid Mech.</u>, 1971, <u>46</u>, 685.

19. F. Gauthier, H.L. Goldsmith and S.G. Mason, <u>Kolloid-Z.</u>,
         1971, <u>248</u>, 1000.

20. E. Anzurowski and S.G. Mason, <u>J.Coll.and Interface Sci.</u>,
      1967, 23, 522; E. Anzurowski and S.G. Mason,
      <u>ibid</u>, p.533; E. Anzurowski, R.G. Cox and S.G. Mason,
      <u>ibid</u>, p.547.

21. D.B. Brooks, J.W. Goodwin and G.V.F. Seaman, <u>J.App.Physiol.</u>,
      1970, <u>28</u>, 172.

22. J.W. Goodwin, <u>Trans.Brit.Ceram.Soc.</u>, 1971, <u>70</u>, 65.

23. B.E. Conway and A. Dobry-Duclaux, in "Rheology, Theory and
      Applications", Vol.3, ed. F. Eirich, Academic Press,
      NY, 1960, p.83.

24. F. Booth, <u>Proc.Roy.Soc.</u>, 1950, <u>A205</u>, 533.

25. B. Krasny-Ergen, <u>Kolloid-Z.</u>, 1936, <u>74</u>, 172.

26. W.B. Russel, <u>J.Fluid Mech.</u>, 1978, <u>85</u>, 673.

27. D.A. Lever, <u>J.Fluid Mech.</u>, 1979, <u>92</u>, 421.

28. F.S. Chan and D.A.I. Goring, <u>J.Coll.and Interface Sci.</u>,
      1966, <u>22</u>, 371; F.S. Chan, J. Blachford and D.A.I. Goring,
      <u>ibid</u>, 378; J. Blachford, F.S. Chan and D.A.I. Goring,
      <u>J.Phys.Chem.</u>, 1969, <u>73</u>, 1062.

29. W.B. Russel, <u>J.Fluid Mech.</u>, 1978, <u>85</u>, 209.

30. J. Stone-Masui and A. Watillon, <u>J.Coll.and Interface Sci.</u>,
      1968, 28, 187.

31. a) H.L. Goldsmith and S.G. Mason, <u>Proc.Roy.Soc.</u>, 1964, <u>6</u>,
         273.

     b) C.L. Darabaner and S.G. Mason, <u>Rheol.Acta</u>, 1967, <u>6</u>, 273.

32. R.C. Ball and P. Richmond, <u>Phys.Chem.Liq.</u>, 1980, <u>9</u>, 99.

33. R. Evans, J.B. Davison and D.H. Napper, <u>Polymer Lett.</u>,
      1972, <u>10</u>, 449.

34. R. Buscall, <u>J.Chem.Soc., Farad.Trans.1</u>, 1981, <u>77</u>, 909.

35. I.M. Krieger and M. Eguiluz, <u>Trans.Soc.Rheol.</u>, 1976, <u>20</u>, 29.

36. W. Luck, M. Klier and H. Wesslau, <u>Naturwissenschaften</u>, 1963,
      <u>50</u>, 485.

37. P.A. Hiltner and I.M. Krieger, <u>J.Phys.Chem.</u>, 1969, <u>73</u>, 2386.

38. S. Hachisu, Y. Kobayashi and A. Kose, <u>J.Coll.and Interface
      Sci.</u>, 1973, <u>42</u>, 343.

39. J.W. Goodwin, R.H. Ottewill and A. Parentich, <u>J.Phys.Chem.</u>,
      1980, 84, 1580.

40. J.C. Brown, P.N. Pusey, J.W. Goodwin and R.H. Ottewill,
      <u>J.Phys.A:Gen.Math.</u>, 1975, <u>8</u>, 664.

41. J.W. Goodwin and R.W. Smith, <u>Disc.Farad.Soc.</u>, 1974, <u>57</u>, 126.

42. J.W. Goodwin and A.M. Khidher, in "Colloid and Interface
      Science, Vol.IV, ed. M. Kerker, Academic Press, NY, p.529.

43. R. Buscall, J.W. Goodwin, M.W. Hawkins and R.H. Ottewill, <u>J.Chem.Soc.,Farad.Trans.I</u>, submitted for publication.

44. L. Barclay, A. Harrington and R.H. Ottewill, <u>Kolloid Z.u.Z. Polymere</u>, 1972, <u>250</u>, 655.

45. R.S. Crandall and R. Williams, <u>Science</u>, 1977, <u>198</u>, 293.

46. S. Mitaku, T. Ohitsuki, K. Evari, A. Kishimoto and K.O. Kano, <u>Japan. J.Appl.Phys.</u>, 1978, <u>17</u>, 305.

47. E. Dubois-Violette, P. Pieranski, F. Rothen and L. Strzelecki, <u>J.Physique</u>, 1980, <u>41</u>, 369.

48. W.J. van Megen, I.K. Snook and R.O. Watts, <u>J.Coll.and Interface Sci.</u>, 1980, <u>77</u>, 131.

49. J.W. Goodwin, T. Gregory and J.A. Stile, <u>J.Coll.and Interface Sci.</u>, submitted for publication.

50. S. Glasstone, K.J. Laidler and H. Eyring in "The Theory of Rate Processes", McGraw-Hill, NY. 1941.

51. T.G.M. van de Ven and S.G. Mason, <u>J.Coll.and Interface Sci.</u>, 1976, <u>57</u>, 505.

52. T.G.M. van de Ven and S.G. Mason, <u>J.Coll.and Interface Sci.</u>, 1976, <u>57</u>, 517.

53. T.G.M. van de Ven and S.G. Mason, <u>Coll.and Polymer Sci.</u>, 1977, <u>255</u>, 468.

54. T.G.M. van de Ven and S.G. Mason, <u>Coll.and Polymer Sci.</u>, 1977, <u>255</u>, 794.

55. G.R. Zeichner and W.R. Schowalter, <u>A.I.Chem.Eng.J.</u>, 1977, <u>23</u>, 243.

56. G.R. Zeichner and W.R. Schowalter, <u>J.Coll.and Interface Sci.</u>, 1979, <u>71</u>, 237.

57. P.A. Reynolds, Ph.D. thesis, Bristol 1980.

58. A.S. Michaels and J.C. Bolger, <u>Ind.Eng.Chem.(Fun.)</u>, 1962, <u>1</u>, 153.

59. B.A. Firth and R.J. Hunter, <u>J.Coll.and Interface Sci.</u>, 1976, <u>57</u>, 266.

60. T.G.M. van de Ven and R.J. Hunter, <u>Rheol.Acta</u>, 1977, <u>16</u>, 534.

61. R.J. Hunter and J. Frayne, <u>J.Coll.and Interface Sci.</u>, 1980, <u>76</u>, 107.

9
Concentrated Dispersions

By R. H. Ottewill

DEPARTMENT OF PHYSICAL CHEMISTRY, SCHOOL OF CHEMISTRY, UNIVERSITY OF BRISTOL, CANTOCK'S CLOSE, BRISTOL BS8 1TS, U.K.

Introduction

In a typical sol with a low number concentration and an intermediate electrolyte concentration the particles can move around freely in the large volume of the dispersion medium. This motion, frequently called Brownian motion, involves the translational diffusion of the particles. It was examined in some detail experimentally by Perrin[1] and theoretically by Einstein[2] in the early part of the twentieth century. In these dilute dispersions only occasional contacts occur between the particles and hence the diffusional motion is only very marginally restricted.

However, as the particle number concentration of the sol or dispersion increases the volume of space occupied by the particles, within the total volume, increases and thus a proportion of the space is excluded in terms of its possible occupancy by any single particle. Moreover, the probability of interactive contact between the particles increases. It is in this situation that the forces between the particles play an important role in determining the overall properties of the dispersion. It has become customary, and useful, to discuss the various interactions which can occur between particles in terms of the energy of interaction between a pair of particles as a function of the distance of separation between the centres of the particles. For the purposes of this article we will consider only spherical particles having a radius R and separated by a distance, r, as shown in Figure 1.

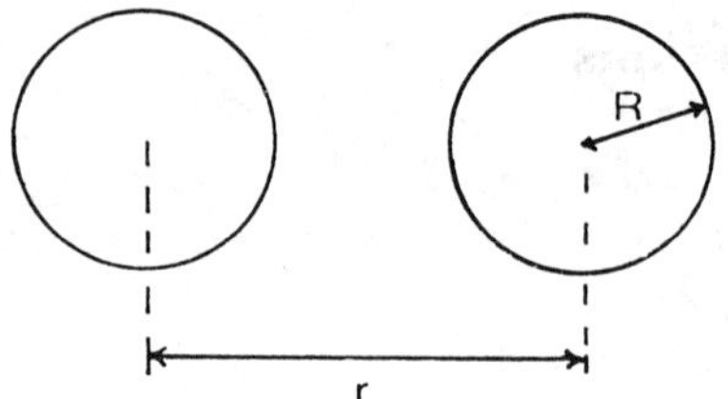

Figure 1:   Interaction between two spheres of radius, R, with
             their centres separated by a distance, r.

The actual volume of a particle in this situation, Vp, is
thus given by

$$Vp = \frac{4}{3} \pi R^3$$

and the total volume of particles in the system ν by

$$\nu = NpVp$$

If the total volume of dispersion taken is V then the actual
volume fraction is defined by

$$\phi = \frac{\nu}{V}$$

and if $V = 1$ cm^3 then $\phi = NpVp$.

Types of Particle-Particle Interaction

1.   Hard-Sphere Interaction.   The simplest interaction between
particles is that of the form shown in Figure 2 in which at a
certain distance of separation, the energy of interaction rises
very steeply to plus infinity[3].   This distance can be defined as

$$r = 2 R_{HS}$$

where R_{HS} in general is greater than R, but in some cases R could
be close to R_{HS}.   We can therefore immediately define a new volume
fraction, the hard sphere volume fraction, ϕ_{HS}, which is given by

$$\phi_{HS} = \frac{\nu_{HS}}{V}$$

with $$\nu_{HS} = Np \cdot \frac{4}{3} \pi R_{HS}^3$$

It follows, since R_{HS} is in general greater than R, that $\phi_{HS} > \phi$.

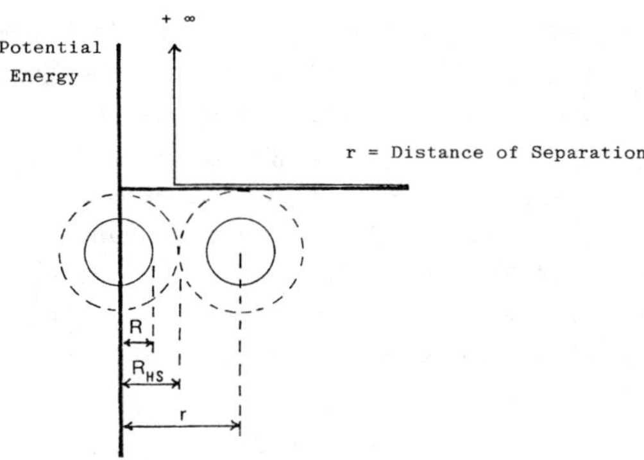

Figure 2:   Interaction between Hard Spheres

2.   <u>Electrical Double Layer Repulsion</u>. In aqueous media the
majority of colloidal particles have an electrical charge on the
surface which gives rise to a surface electrostatic potential, ψ_s,
and to a potential energy of electrostatic repulsion which for
small spherical particles can be given by the expression[4]

$$V_R = \frac{4\pi\varepsilon_r\varepsilon_o R^2 \psi_s^2 \cdot \exp\left[-\kappa(r - 2R)\right]}{r} \qquad \ldots \quad (1)$$

with ε_r = the relative permittivity of the medium ε_o = the
permittivity of a vacuum and κ = the reciprocal Debye-Hückel
electrical double layer thickness. The latter quantity is related
directly to the concentration of a symmetrical electrolyte, c,
expressed in mol dm^{-3} by

$$\kappa^2 = \frac{2 \, v^2 e^2 \, N_A \, c}{\varepsilon_r\varepsilon_o \, kT} \qquad \ldots \quad (2)$$

with v = valency of the ions, e = the fundamental electronic
charge, k = Boltzmann's constant, T = absolute temperature and
N_A = Avogadro's number.

From equation (1) it is clear that the change of interaction
energy with r is dependent on the term $\exp\left[-\kappa(r - 2R)\right]/r$. Consequently the rate of decrease of repulsion with distance is
strongly dependent on κ. At low electrolyte concentrations therefore (e.g. 10^{-5} mol dm^{-3}) electrostatic repulsion provides a long
range term whereas at high electrolyte concentrations the repulsive energy term decays very rapidly. The form of the curve of V_R

against r is shown schematically in Figure 3. Since the force of
repulsion between the particles is determined by the gradient of
the curve it can be seen that at a certain distance of approach
the repulsive force can become so strong that closer approach is
improbable. This provides a useful concept (see later) since
again the particle can be assigned an effective radius and as
a working hypothesis the concept of a hard-sphere potential with
a soft tail can be utilised. In essence this means that the
particle can behave as a hard sphere with an effective radius
much larger than its actual radius.

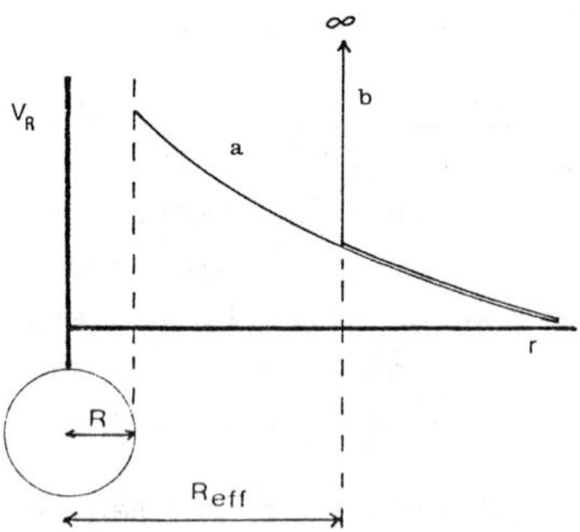

Figure 3:   Repulsive interaction between charged spheres
            a) V_R from equation (1);   b) Hard Sphere potential
            with a "soft" tail

3.   <u>van der Waals Attraction</u>. The attractive energy of interaction, V_A, between two spherical particles of equal radius was
originally given by Hamaker[5] in the form

$$V_A = - \frac{A}{12}\left\{ \frac{1}{x^2 + 2x} + \frac{1}{x^2 + 2x + 1} + 2\ln\frac{x^2 + 2x}{x^2 + 2x + 1} \right\} \ldots \quad (3)$$

where $x = (r - 2R)/2R$ and A is the composite Hamaker constant for
the particles in the medium, as given by

$$A = (A_{11}^{\frac{1}{2}} - A_{22}^{\frac{1}{2}})^2 \qquad \ldots \quad (4)$$

with A_{11} = the Hamaker constant of the particles and A_{22} that of

the medium. The form of the function given by equation (3) is
shown schematically in Figure 4; more detailed accounts are given
elsewhere[6]. In the case of attractive interactions as soon as the
energy of attraction becomes greater than that of the Brownian
kinetic energy of the particles attraction must inevitably result
with the subsequent formation of doublets, triplets and eventually
higher multiplets[7].

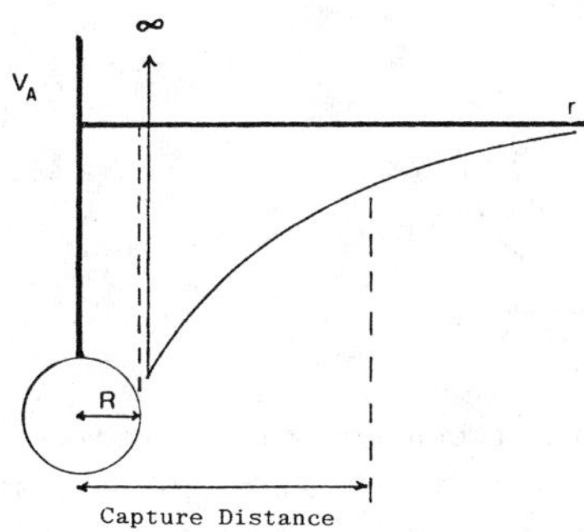

Figure 4:   Attractive interaction between two spheres according
            to equation (3) with a very short range strong repul-
            sion ($R \sim R_{eff}$)

4.   <u>Steric Interactions</u>. The topic of steric stabilisation has
been dealt with in detail earlier in this volume by Napper[8] and
it is unnecessary to reiterate here the many equations for the
energy of steric interaction energies which have been postulated
in the literature. It is clear from the form of these that when
steric repulsion occurs it can only begin to occur at the distance
of centre to centre separation where overlap of the adsorbed
layers commences. Since many of the details of this process are
still obscure, for the purposes of this article we will assume that,
as a first approximation, it can be treated as a hard sphere
interaction with an R_{eff} roughly equal to that of the particle
core of radius R plus the thickness of the adsorbed layer, δ,
as shown in Figure 5.

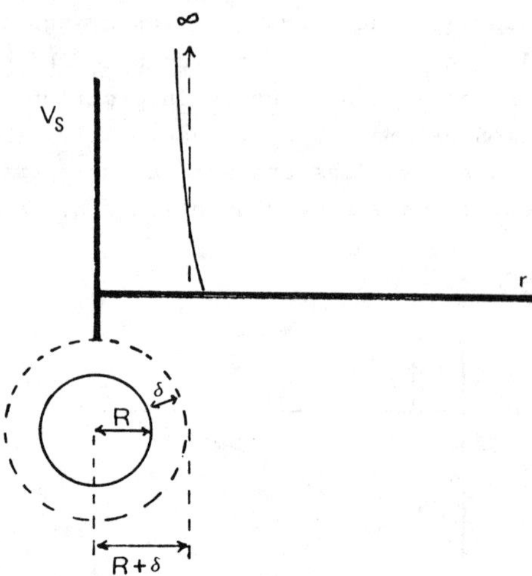

Figure 5:   Interaction between spheres of radius R, with
            adsorbed layers, thickness δ; ———, steric repulsion
            model; — — —, Hard Sphere model

5.   <u>Combinations of Interactions</u>. In practice it is possible to
obtain colloidal dispersions in which various combinations of
electrostatic, attractive and steric interactions occur. A full
discussion is not appropriate here and the subsequent discussion
will largely be dominated by situations in which only electrostatic or steric repulsions occur under well-defined conditions.

<u>Concentrated Systems</u>
     The idea of pair potentials of various types elaborated in
the last section must now be extended into the context of
concentrated systems. If a very simplistic view is taken we can
obtain a schematic picture of the situation as that shown in
Figures 6a and 6b. In a situation where the repulsive forces are
dominant there is a strong dependence on the range of the repulsion, R_{eff}, and as more particles are added each particle must
interact with others with the consequent formation of a dispersion
with a high degree of order. In this system each particle will
have a high co-ordination number, which could approach 12 on say
the formation of a face-centred cubic array. It becomes very
important to recognize that in such a system the particles can be
well separated by the dispersion medium and that particularly in

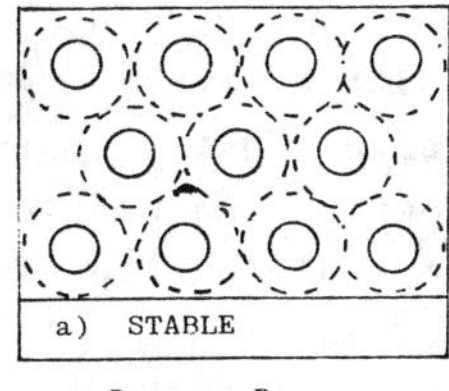
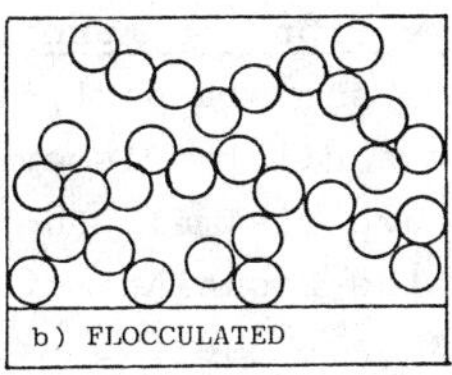

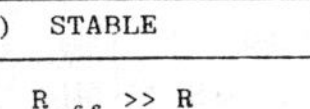

Figure 6:   Comparison of particle arrangement in, a) a stable
            system, b) a flocculated system.
            ———— , particle radius R; – – –, R_{eff}.

the case of aqueous dispersions at low electrolyte concentrations
R_{eff} can be many times R so that these effects can be observed
in dispersions at low volume fractions. From a fundamental point
of view our interests must lie in determining the spatial and
temporal correlations of the particles in such a system. In the
former case, by analogy to states of matter, we must ask is the
arrangement of the particles in the dispersion "vapour-like"
(random), "liquid-like" (loosely ordered) or "solid-like" (highly
ordered), and in the case of temporal correlations we must ask how
the diffusion of a single particle is modified under conditions
where it is subjected to many body interactions[9]. Ultimately,
once the system is understood at the microparticulate level we
must understand how the particle-particle interactions influence
the macroscopic properties of the dispersion, such as its osmotic
pressure, its elasticity and its flow properties.

     In a concentrated dispersion once the attractive interactions
become the dominant ones the particles will "stick" on coming into
an attractive energy well with the consequent formation of a
highly disordered system of the type shown schematically in
Figure 6b. In such a system there is a high void volume and each
particle now has a low co-ordination number, e.g. 2 or 3. The
spatial correlations have essentially become non-existent and the
temporal correlations are largely determined by the behaviour of
the whole network. The latter factor makes this type of system
particularly appropriate for study by rheological techniques of
the type discussed by Goodwin[10].

The Radial Distribution Function

Figure 6a shows a particle distribution with a high degree of order. In a dilute dispersion in which the interparticle interactions are minimal, the arrangement of the particles will be as random as the Brownian motion of the individual particles allows. Between these two extremes a range of other possibilities exists and a method is needed by which the state of the system can be expressed. This can be expressed by reference to Figure 7. In a container of volume V containing Np particles the average macroscopic density can be expressed as

$$\rho_o = \frac{Np}{V} \qquad \qquad \ldots \quad (5)$$

However, if we take a microscopic view of the system as seen by one particular particle then a somewhat different picture emerges. As one proceeds outwards from the particle at the origin it is immediately clear that there is space surrounding it and so the density of particles in that region is zero. On increasing the distance r from the centre of the particle, however, and examining the first shell of thickness dr it can be seen that there are four particles in the two-dimensional perspective and therefore even more on a three-dimensional basis. As r becomes very large, however, the number density of particles in a shell of thickness dr must increase and approach ρ_o. Thus, we can define a number density function $\rho(r)$ which is a function of r and which, for r just marginally greater than 2R must be zero and for $r \to \infty$ must be equal to ρ_o.

Thus a distribution function

$$g(r) = \frac{\rho(r)}{\rho_o} \qquad \qquad \ldots \quad (6)$$

will similarly commence at zero for $r \simeq 2R$ and become unity as $r \to \infty$. From the schematic picture of Figure 7 it can be seen that for a highly ordered arrangement, e.g. a crystalline array, g(r) will have a series of spikes as a function of r whereas for a less regular arrangement g(r) will be an oscillating function of r. It can also be stated that the exclusion of particles around the particle at the origin will depend on the type of interaction involved and hence at short distances g(r) will be related to the interaction energy.

The term RADIAL DISTRIBUTION FUNCTION is usually used for the quantity

$$4 \pi r^2 \rho(r)$$

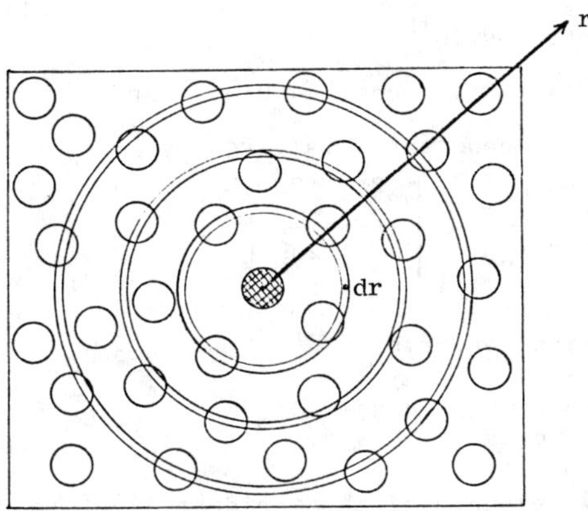

Figure 7:   Schematic illustration of the radial distribution
            of particles around a central particle

although it is also used by some authors for the term g(r).

   Both $\rho(r)$ and g(r) are useful quantities to obtain for an
interacting system since they immediately provide information
about the spatial distribution of the particles in the dispersion.
The most appropriate way of determining this quantity is from
scattering experiments and the method of approach will now be
outlined.

The Scattering of Radiation from Non-Interacting Particles

   The intensity of scattering as a function of the scattering
vector Q from an assembly of Np spherical particles relative to
the intensity of the incident radiation is given by[11]

$$I(Q) = A. \left[\text{Material Factor}\right] . NpVp^2. P(Q) \qquad \ldots \quad (7)$$

where A = an instrument constant, Vp = the volume of each spherical
particle ($4\pi R^3/3$) and P(Q) = the particle form factor for a sphere
of radius R given by

$$P(Q) = \left[3(\sin QR - QR . \cos QR)/(QR)^3\right]^2 \qquad \ldots \quad (8)$$

The material factor is dependent on the nature of the radiation
employed for the experiment. In the case of light scattering it
is dependent on the refractive index of the particle, n_p, and
the refractive index of the medium, n_m. The appropriate equation
for the intensity of light scattered by an assembly of Np

independent particles is[12]

$$I(Q) = A.K. (1 + \cos^2 \theta) P(Q) \cdot N_p V_p^2 \qquad \ldots \qquad (9)$$

where the material constant K is given by

$$K = \frac{9\pi^2 n_m^4}{2 \lambda_o^4} \left[\frac{n_p^2 - n_m^2}{n_p^2 + 2n_m^2} \right]^2 \qquad \ldots \qquad (10)$$

and the scattering vector by

$$Q = \frac{4\pi n_m}{\lambda_o} \sin(\theta/2) \qquad \ldots \qquad (11)$$

with λ_o = the wavelength of the incident light *in vacuo* and θ = the scattering angle.

In the case of neutron scattering the intensity of scattering is given by

$$I(Q) = A (\rho_p - \rho_m)^2 \cdot P(Q) \cdot N_p V_p^2 \qquad \ldots \qquad (12)$$

with

$$Q = \frac{4\pi}{\lambda} \cdot \sin (\theta/2) \qquad \ldots \qquad (13)$$

where λ = the wavelength of the incident neutron beam, ρ_p = the scattering length density of the particle and ρ_m = the scattering length density of the medium. Some typical data obtained using latex systems have been reported earlier in this volume[13].

<u>The Scattering of Radiation from Particles in Interacting Systems</u>

Once the particles have formed an array of the type illustrated in Figure 7, an allowance has to be made for the fact that owing to some degree of order in the system interference effects occur between the scattered radiation from the particles. In Figure 7 the system was divided into spherical shells of thickness dr. It can be shown that for a single shell of radius r, and thickness dr, the form factor is given by[14]

$$\frac{\sin Qr}{Qr}$$

and hence for a distribution $\rho(r)$ there will be a contribution on integrating over all shells of

$$\int_o^\infty 4\pi r^2 \rho(r) \frac{\sin Qr}{Qr} dr \qquad \ldots \qquad (14)$$

This contribution will arise in addition to that from the individual particles and hence the total scattering arising from such an assembly of particles will be given by

$$I(Q)_{Total} = I(Q)_{particles} \left\{ 1 + \int_0^\infty 4\pi r^2 \, \rho(r) \, \frac{\sin Qr}{Qr} \, dr \right\} \quad \ldots \ (15)$$

where the expression for I(Q) particles is given by equations (9) or (12) according to the type of radiation used for the experiment.

Equation (15) can be rewritten in the form

$$I(Q)_{Total} = I(Q)_{particles} \left\{ 1 + \frac{4\pi}{Q} \int_0^\infty r \left[\rho(r) - \rho_o \right] \sin Qr.dr \right.$$

$$\left. + \int_0^\infty r \, \rho_o \, \sin Qr.dr \right\} \quad \ldots \ (16)$$

The last integral in the curly brackets can be shown to be zero, except at small values of Q, thus giving

$$I(Q)_{Total} = I(Q)_{particles} \left\{ 1 + \frac{4\pi \rho_o}{Q} \int_0^\infty r \left[\frac{\rho(r)}{\rho_o} - 1 \right] \sin Qr.dr \right\} \quad (17)$$

The term in curly brackets in equation (17) is known as the structure factor, S(Q), and remembering that $g(r) = \rho(r)/\rho_o$ this can be written as

$$S(Q) = 1 + \frac{4\pi \rho_o}{Q} \int_0^\infty r \left[g(r) - 1 \right] \sin Qr.dr \quad \ldots \ (18)$$

giving also

$$I(Q)_{Total} = I(Q)_{particles} \cdot S(Q) \quad \ldots \ (19)$$

Thus, in the case of a neutron scattering experiment, by combination with equation (12) we obtain for interacting systems

$$I(Q) = A.NpVp^2 \, (\rho_p - \rho_m)^2 \, P(Q) \, S(Q) \quad \ldots \ (20)$$

Thus S(Q) can be obtained experimentally from scattering measurements on interacting systems. Moreover, Fourier transformation of equation (18) leads directly to

$$g(r) = 1 + \frac{1}{2\pi^2 r\rho_o} \int_0^\infty \left[S(Q) - 1 \right] Q \sin Qr.dQ \quad \ldots \ (21)$$

For non-interacting particles it follows that $S(Q) = g(r) = 1$ and the scattering law is that given by equations (9) and (12).

<u>The Relation between g(r) and the Interaction Potential</u>

In the schematic representation shown in Figure 7 the particles are in Brownian motion and continuously interacting so that for a steady state condition we can equate the diffusional force, $d\mu/dr$, (= the chemical potential gradient) and the interaction force, dV/dr, to give

$$\frac{d\mu}{dr} = \frac{dV}{dr}$$

whence in terms of number density concentration, $\rho(r)$, we obtain

$$kT \; \frac{d \ln \rho(r)}{dr} = \frac{d V(r)}{dr}$$

and

$$kT \int_{\rho_o}^{\rho(r)} d \ln \rho(r) = \int_{Vo}^{Vr} d V(r)$$

Hence it follows that

$$g(r) = \frac{\rho(r)}{\rho_o} = \exp\left(\frac{V(r) - Vo}{kT}\right) \qquad \ldots \quad (22)$$

and that the distribution function g(r) is directly related to the interaction potential. However, (V(r) - Vo) is not the simple pair-potential such as the V_R given earlier. It is more accurately described as the potential of mean force, Φ, such that

$$\Phi = (V(r) - Vo) = V_R + \Psi \qquad \ldots \quad (23)$$

where Ψ is a perturbation occurring as a consequence of the many body interactions. In the limit $r \to 2R$ then $\Phi \to V_R$.

<u>The Osmotic Compressibility</u>

In the model described so far we have considered the microscopic nature of the array of particles. A useful macroscopic property which can be obtained directly is the osmotic compressibility. This is related to the structure factor at $Q = 0$, i.e. $S(0)$ by

$$\left(\frac{\partial \Pi}{\partial \rho_o}\right)_T = \frac{kT}{S(0)} \qquad \ldots \quad (24)$$

S(O) can be determined from scattering experiments by extrapolation of the data to zero Q. Π is the excess osmotic pressure in the system owing to the presence of the particles.

<u>The Relation of S(Q) to Models of Particle-Particle Interaction</u>

The simplest model for interaction as mentioned earlier is that designated as the Hard Sphere model in the physics of simple liquids as proposed by Percus and Yevick[15]. It was shown by Ashcroft and Lekner[16] that this model leads directly to the structure factor, S(Q), in the form

$$S(Q) \;=\; \frac{1}{\left[1 - Np \cdot C(2Q\,R_{HS})\right]} \qquad\qquad \dots \quad (25)$$

·with

$$C(2Q\,R_{HS}) \;=\; -\,32\pi\,R_{HS}^{3} \int_{o}^{1} \frac{\sin(2s\,Q\,R_{HS})}{2s\,Q\,R_{HS}} (\alpha + \beta s + \gamma s^{3})s^{2}\,ds$$

$$\dots \quad (26)$$

where R_{HS} is the radius of the interacting hard sphere such that the hard sphere volume fraction is defined by

$$\phi_{HS} \;=\; \frac{4}{3}\,\pi\,R_{HS}^{3} \cdot Np \qquad\qquad \dots \quad (27)$$

The coefficients α, β and γ are defined by

$$\alpha \;=\; \left.(1 + 2\,\phi_{HS})^{2}\middle/(1 - \phi_{HS})^{4}\right. , \qquad\qquad \dots \quad (28)$$

$$\beta \;=\; \left.-\,6\,\phi_{HS}\,(1 + 0.5\,\phi_{HS})^{2}\middle/(1 - \phi_{HS})^{4}\right. , \quad \dots \quad (29)$$

$$\gamma \;=\; \left.0.5\,\phi_{HS}\,(1 + 2\,\phi_{HS})^{2}\middle/(1 - \phi_{HS})^{4}\right. \qquad \dots \quad (30)$$

Some calculated curves using these equations are given in Figure 8 for particles with R_{HS} = 250 Å at ϕ_{HS} values of 0.25, 0.35 and 0.45.

Expressions for S(Q) based on an electrostatic repulsive potential of the form given in equation (1) have been given by Hayter and Penfold[17] using the mean spherical approximation approach. A comparison of their model with the Hard Sphere model is given in Figure 9.

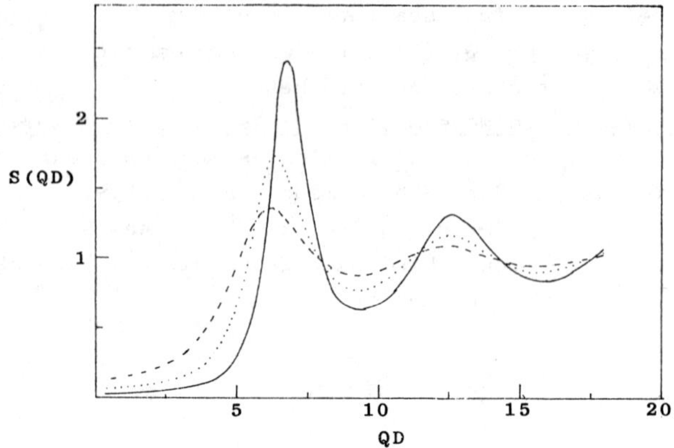

Figure 8:   S(QD) against QD as calculated using equation (25) with R_{HS} = 250 Å and various values of ϕ_{HS}. – – –, ϕ_{HS} = 0.25; ·····, ϕ_{HS} = 0.35; ———, ϕ_{HS} = 0.45, D = 2 R_{HS}.

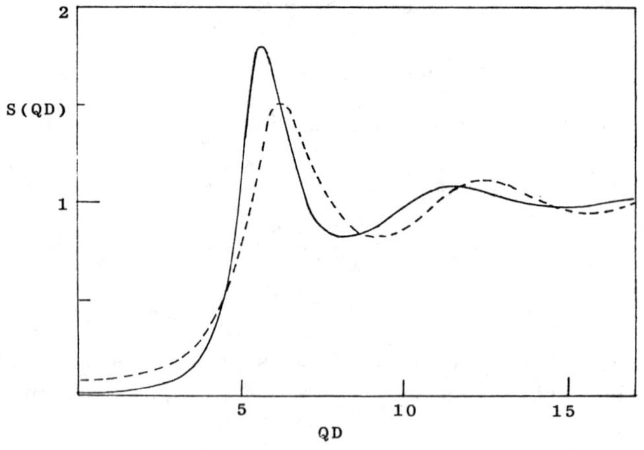

Figure 9:   S(QD) against QD for different models. ——— , Hard Sphere model, ϕ_{HS} = 0.3; – – –, Mean Spherical Approximation with ψ_s = 77 mV, after Hayter and Penfold[17]

Experimental Investigations of S(Q)

Sterically Stabilised Systems. A colloidal dispersion suitable
for small angle neutron scattering studies is that of poly-deuteromethyl methacrylate particles stabilised by poly-12-hydroxystearic
acid and dispersed in a hydrocarbon solvent[18]. A schematic
representation of the nature of the particle is given as Figure 10
earlier in this volume[13]. The form of the curve of I(Q) against
Q for a system with a volume fraction of 0.35 is shown in
Figure 10.

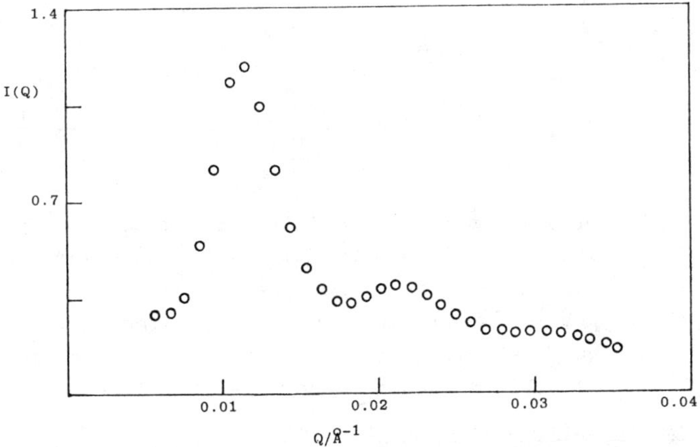

Figure 10: I(Q) in arbitrary units against Q obtained from small
angle neutron scattering with a polydeutero methyl
methacrylate latex, stabilised by poly-12-hydroxy-
stearic acid, in d_{18}-octane. Effective volume fraction = 0.35.

The resulting curve of S(Q) vs Q obtained from the same data
is shown in Figure 11. In this Figure the experimental results
are compared with the calculations carried out using equations (25)
to (30) based on a hard sphere model with an effective hard sphere
volume fraction of 0.35 and a hard sphere radius of 280 Å.

A good fit is obtained between the experimental results and
the theoretical calculations for this sytem, indicating that the
basic hard sphere model gives a good representation of the data.
This seems to fit with the basic idea of steric stabilisation for
a system of this type, which has an adsorbed layer composed of

grafted chains of approximately equivalent length, thus
essentially forming a shell on the outside of the particle.

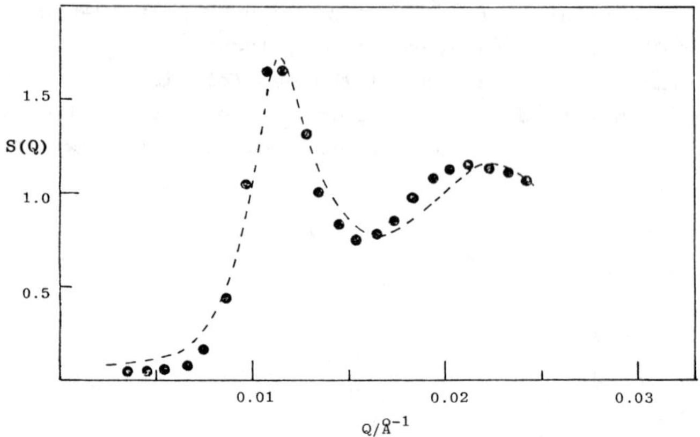

Figure 11:   S(Q) against Q for a polydeuteromethyl methacrylate
             latex, stabilised by poly-12-hydroxystearic acid, in
             d_{18}-octane. ● , experimental data from small angle
             neutron scattering;— — , calculated curve using a
             Percus-Yevick model with an effective volume fraction
             of 0.35 and a hard sphere radius of 280 Å.

<u>Electrostatically Stabilised Systems</u>

      Figure 12 shows the form of the curve of S(Q) against Q for
a polystyrene latex of actual volume fraction, 0.138, containing
particle diameter = 360 Å. On the same figure the calculated
curve is given for an effective volume fraction, ϕ_{HS} = 0.466
obtained using the Percus-Yevick approach. This corresponds to a
hard sphere radius for the particles, R_{HS} = 255 Å. Hence, under
these conditions the ratio R_{HS}/R, with R = the actual particle
radius, is 1.42. It is of considerable interest that the experimental results and hard sphere computations are in reasonably good
agreement. This suggests that the interaction conditions are close
to those represented in Figure 3, namely, that of a hard sphere
potential with a soft tail implying that once the particles
approach to within a certain distance the repulsion is so strong
that essentially the particles behave as hard spheres. Thus the
value of R_{eff} gives an indication of the range of effective double
layer repulsion.

The data of Figure 12 can be transposed to the radial distribution function $4\pi r^2 \rho(r)$ as a function of r and this is shown in Figure 13.

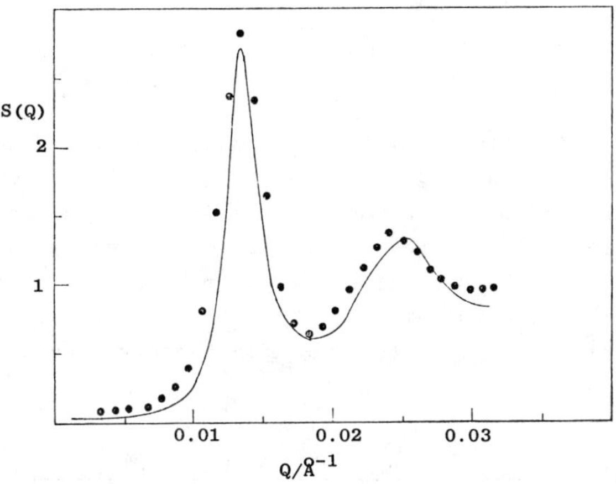

Figure 12: S(Q) against Q for an electrostatically stabilised polystyrene latex (R = 180 Å; ϕ = 0.138) in water. ● , experimental points from small angle neutron scattering; ———, Hard Sphere model using $_{eff}$ = 0.47, R_{HS} = 255 Å.

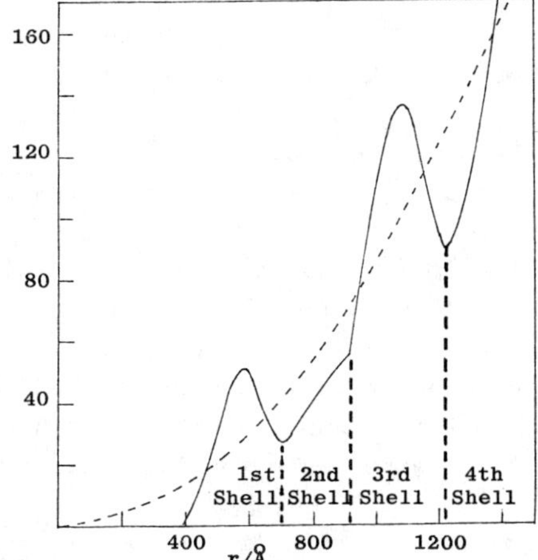

Figure 13: Radial distribution function plotted against distance of separation, r. ———, $4\pi r^2 g(r)\, \rho_o$; — — —, $4\pi r^2 \rho_o$.

Plotted in this form the results show clearly the formation of shells of particles around the central particle. The co-ordination number of each shell can be obtained by integration, that is,

$$n = \int_{r_n}^{r_m} 4\pi r^2 \, \rho(r) \, dr$$

where r_m and r_n gives the limits of the shell. For the nearest shell to the central particle we find a co-ordination number of 10.3 and a mean radius for the shell r_1 = 58 nm. A second shell occurs at r = 820 Å, equivalent to $r_1\sqrt{2}$, with a co-ordination number of 6.3. The third shell occurs at r = 1080 Å, equivalent to $r_1\sqrt{3.5}$, with a co-ordination number of 30. These results are compared with those expected for a face-centred cubic packing in Table 1.

Table 1

Structural Features of a Concentrated Latex Dispersion

	Experimental		f.c.c. Packing of Spheres	
Shell	Co-ordination Number	r	Co-ordination Number	r
1	10.3	r_1	12	r_1
2	6.3	$r_1\sqrt{2}$	6	$r_1\sqrt{2}$
3	30.0	$r_1\sqrt{3.5}$	24	$r_1\sqrt{3}$

These results show that as a consequence of the electro-static repulsion between the particles the structure in the system is approaching that of face-centred-cubic packing. However, the broad peaks and the lower co-ordination number of the first peak indicate that the structure still appears to have some "liquid-like" character.

For this system it is also possible to calculate the potential of mean force as a function of interparticle centre to centre separation. The results are shown in Figure 14. The oscillatory nature found shows that many body effects are import-ant in systems of this type and that the particles in the second and third shells still experience repulsive forces from the central particle.

The above results are taken from small angle neutron scattering experiments carried out on systems containing a high number concentration of particles. It becomes essential under

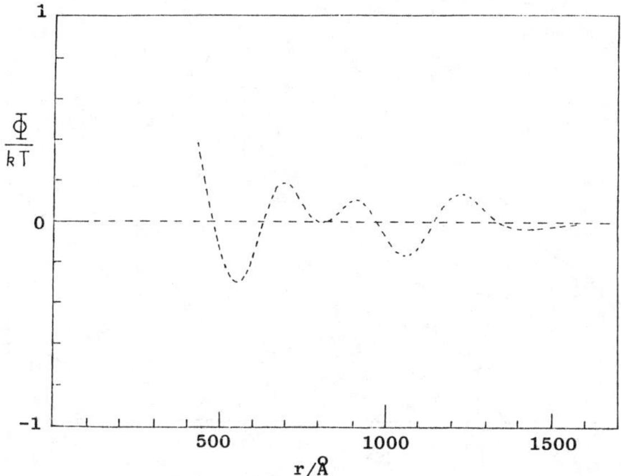

Figure 14:   Φ/kT against distance of separation for an electrostatically stabilised polystyrene latex (R = 180 Å;
φ = 0.138) in water.

these conditions to use neutron beams in order to avoid the
multiple scattering effects which would occur with transmitted
light.   However, because of the very long range of electrostatic
repulsive forces interaction effects can be observed in very
dilute latex systems at very low electrolyte concentrations[19].
If small diameter particles are used (ca. 50 nm) angular light
scattering measurements can be made on dilute dispersions in
order to obtain S(Q) as a function of Q and hence S(O) obtained
by extrapolation.   If measurements are made also as a function of
number concentration then integration using

$$\Pi \;=\; kT \int_{0}^{\rho_{0}} \frac{1}{S(O)}\, d\rho$$

provides the excess osmotic pressure, Π, as a function of
concentration.   Some results obtained on a polystyrene latex
containing spherical particles of radius 156 Å[20] are shown in
Figures 15a and 15b.

The excess osmotic pressures obtained by scattering experiments can then be compared with those obtained directly by
compression measurements.

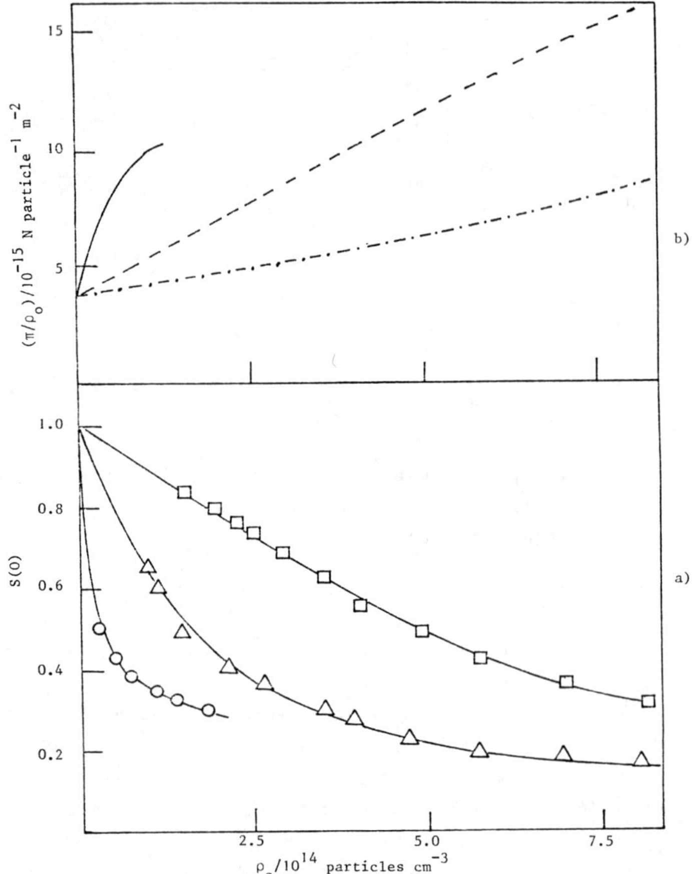

Figure 15: a) S(O) against ρ_o; —O—, ion-exchanged system;
—△—, 10^{-4} mol dm^{-3} sodium chloride;
—□—, 5×10^{-3} mol dm^{-3} sodium chloride.
b) Derived π/ρ_o against ρ_o from S(O) against ρ_o data;
———, ion-exchanged system; — — —, 10^{-4} mol dm^{-3} sodium chloride; —·—·, 5×10^{-3} mol dm^{-3} sodium chloride.
Polystyrene latex with R = 156 Å.

References

1.   M. Jean Perrin, "Brownian Movement and Molecular Reality",
         Taylor and Francis, London, 1910.

2.   A. Einstein, "Investigations on the Theory of Brownian
         Motion" (English Trans.), Methuen, London, 1926.

3.   D.A. McQuarrie, "Statistical Thermodynamics", Harper and
         Row, London, 1973, p.307.

4.   E.J.W. Verwey and J.Th.G. Overbeek, "Theory of the Stability
         of Lyophobic Colloids", Elsevier, Amsterdam, 1948.

5.   H.C. Hamaker, _Physica_, 1937, _4_, 1058.

6.   H.R. Kruyt, "Colloid Science", Elsevier, Amsterdam, 1952,
         vol.1.

7.   M. von Smoluchowski, _Z.Phys.Chem._, 1917, _92_, 129.

8.   D.H. Napper, this volume, Chapter 5.

9.   P.N. Pusey and R.A. Tough, "Dynamic Light Scattering and
         Velocimetry", Ed. R. Pecora, Plenum, New York, 1982,
         in press.

10.   J.W. Goodwin, this volume, Chapter 8.

11.   D.J. Cebula, D. Myers and R.H. Ottewill, _Colloid and Polymer
         Science_, 1982, in press.

12.   M. Kerker, "The Scattering of Light and Other Electromagnetic
         Radiation", Academic Press, New York, 1969.

13.   R.H. Ottewill, this volume, Chapter 7.

14.   H.C. van de Hulst, "Light Scattering by Small Particles",
         John Wiley and Sons, New York, 1957.

15.   J.K. Percus and G.J. Yevick, _Phys.Rev._, 1958, _110_, 1.

16.   N.W. Ashcroft and J. Lekner, _Phys.Rev._, 1966, _45_, 33.

17.   J.B. Hayter and J. Penfold, _Molecular Physics_, 1981, _42_, 109.

18.   R.J.R. Cairns, R.H. Ottewill, D.W.J. Osmond and I. Wagstaff,
         J.Colloid and Interface Science, 1976, _54_, 45.

19.   J.C. Brown, P.N. Pusey, J.W. Goodwin and R.H. Ottewill,
         J.Phys.A., 1975, _8_, 664.

20.   R.H. Ottewill and R.A. Richardson, _Colloid and Polymer
         Science_, 1982, in press.